29th International Conference of the International Nuclear Target Development Society (INTDS2018)

EPJ Web of Conferences Volume 229 (2020)

East Lansing, Michigan, USA
8 - 12 October 2018

Editors:

Frederique Pellemoine
Wolfgang Mittig
Christelle Stodel

ISBN: 978-1-7138-0827-5

Printed from e-media with permission by:

Curran Associates, Inc.
57 Morehouse Lane
Red Hook, NY 12571

Some format issues inherent in the e-media version may also appear in this print version.

No changes have been made to the content of these proceedings. There may be changes to pagination and minor adjustments for aesthetics.

Printed with permission by Curran Associates, Inc. (2021)

For additional information, please contact EDP Sciences – Web of Conferences
at the address below.

EDP Sciences – Web of Conferences
17, Avenue du Hoggar
Parc d'Activité de Courtabœuf
BP 112
F-91944 Les Ulis Cedex A
France

Phone: +33 (0) 1 69 18 75 75
Fax: +33 (0) 1 69 28 84 91

contact-edps@webofconferences.org

Additional copies of this publication are available from:

Curran Associates, Inc.
57 Morehouse Lane
Red Hook, NY 12571 USA
Phone: 845-758-0400
Fax: 845-758-2633
Email: curran@proceedings.com
Web: www.proceedings.com

TABLE OF CONTENTS

29th International Conference of the International Nuclear Target Development Society (INTDS2018)

Frederique Pellemoine*[,1], *Wolfgang* Mittig[2], and *Christelle* Stodel[3]

[1] Michigan State University, Facility for Rare Isotope Beams, MI 48824, USA
[2] Michigan State University, National Superconducting Cyclotron Lab, MI 48824, USA
[3] Grand Accélérateur National d'Ions Lourds (GANIL), CNRS/CEA, Bvd Henri Becquerel, BP 55027, 14076 Caen CEDEX 5, France

The 29th International Conference of International Nuclear Target Development Society (INTDS2018) was hosted by the Facility for Rare Isotope Beams (FRIB) held on 8-12 October 2018 on the campus of Michigan State University in East Lansing, Michigan, USA.

The series of INTDS conferences dates back to 1963 when the "first Symposium on Research Materials for Nuclear Measurements (CBNM, recently IRMM – Institut for Reference Materials and Measurements) in Belgium. Since then the society shares experience and discusses the problems related to the target preparation during conferences (recently biennial) hosted alternatively mainly in Europe, North America and Asia (www.intds.org).

Fifty participating scientists from 13 countries (Belgium, Canada, China, Finland, France, Germany, Japan, Poland, Romania, South Africa, Switzerland, UK, USA), met in East Lansing presenting over 45 scientific contributions in seven different sessions titled: *beam charge strippers (foil, liquid, gas, plasma), target characterization, thin films and foils preparation techniques, isotopically enriched and radioactive targets, targets for high intensity beams, cryogenic and polarized targets, targets for special application (medical, industrial, controlled fusion).*

A laboratory tour of the Facility for Rare Isotope beams was proposed in the program including the new FRIB accelerator under construction as well as the challenging equipment such as the liquid lithium stripper, the production target and the beam dump assembly. The Wednesday afternoon was dedicated to visiting Henry Ford and Greenfield Village in the neighbourhood of Detroit. The day ended with the night conference dinner on a cruise ship, the Detroit Princess boat.

We sincerely appreciated the cooperative contribution of all members of the Local Organizing Committee for their efficient management at the conference. We wish to warmly thank the members of the Scientific Advisory Committee for helping us with the selection of the speakers and precious suggestions.

We also express our gratitude to the INTDS review committee for their careful reading of papers making possible their publication in this EPJ issue.

Scientific Advisory Committee

- K. Eberhardt, Johannes Gutenberg-Universitat Mainz & HI Mainz, Germany
- D. Giliam, NIST, Washington, USA
- J. Greene, ANL, Argonne, USA
- N. Kheswa, iThemba LABS, Faure, South Africa
- B. Kindler, GSI, Darmstadt, Germany
- H. Kuboki, RIKEN RNC, Wako, Japan
- B. Lommel, GSI, Darmstadt, Germany
- G. Sibbens, IRMM, Geel, Belgium
-
- D. Steski, BNL, Upton NY, USA
- C. Stodel, GANIL, Caen, France
- A. Stolarz, HIL-UW, Warsaw, Poland
- Y. Yamazaki, JAEA, Chiba, Japan
- S. Zeisler, TRIUMF, Vancouver, Canada

Local Organizing Committee

- F. Pellemoine (chair), FRIB, East Lansing, USA
- W. Mittig (co-chair), NSCL, East Lansing, USA
- A. Lesage, FRIB, East Lansing, USA
- A. McCausey, FRIB, East Lansing, USA

Guest editors

- F. Pellemoine, FRIB, East Lansing, USA
- W. Mittig, NSCL, East Lansing, USA
- C. Stodel, GANIL, Caen, France

Sponsors

- American Elements
- ACF METALS
- Micromatter Technologies Inc.
- Applied Nanotech, Inc.

List of participants

Name	Email	Institution
Dr. AYYAD, Yassid	yassid.ayyad@gmail.com	LBNL, USA
Mr. BARBER, Powell	pbarber@fsu.edu	Florida State University
Dr. BHANDARKAR, Suhas	bhandarkar1@llnl.gov	LLNL, USA
Dr. BOND, Evelyn	bond@lanl.gov	LANL, USA
Dr. BREDEWEG, Todd	toddb@lanl.gov	LANL, USA
Dr. COPPING, Roy	coppingr@ornl.gov	ORNL, USA
Dr. DRESSLER, Rugard	rugard.dressler@psi.ch	PSI, Switzerland
Dr. ESKER, Nicholas	neesker@triumf.ca	TRIUMF, Canada
Dr. FINK, Richard	DFink@appliednanotech.net	Applied Nanotech Inc., USA
FITSOS, Pete	fitsos1@llnl.gov	LLNL, USA
Dr. GILLIBERT, Alain	alain.gillibert@cea.fr	CEA/IRFU/DPhN, France
Dr. GOTT, Matthew	mgott@anl.gov	ANL, USA
Mr. GREENE, John	greene@anl.gov	ANL, USA
Mr. HASEBE, Hiroo	hasebe@riken.jp	RIKEN, Japan
Dr. HE, Xiaoshan	heaxiaoashan@163.com	Research Center of Laser Fusion, China
Ms. JIANG, Wen	jiang568@purdue.edu	Purdue University, USA
Dr. KANEMURA, Takuji	kanemura@frib.msu.edu	FRIB, USA
Mrs. KHESWA, ntombizonke	kheswa@tlabs.ac.za	iThemba LABS, South Africa
Dr. KINDLER, Birgit	b.kindler@gsi.de	GSI, Germany
Dr. KUBOKI, Hironori	kuboki@post.j-parc.jp	J-PARC (KEK), Japan
Dr. KUNZ, Peter	pkunz@triumf.ca	TRIUMF, Canada
Mr. LIANG, Johnson	liangj27@mcmaster.ca	McMaster University, Canada
Dr. LIU, Yansong	jdliuyansong@163.com	Research Centre for Laser Fusion, China
Dr. LI, Xibo	xiboli@caep.cn	Science and Technology on Plasma Physics Laboratory, Research Center of Laser Fusion, China
Dr. LOMMEL, Bettina	b.lommel@gsi.de	GSI , Germany
Dr. MAUGERI, Emilio Andrea	emilio-andrea.maugeri@psi.ch	PSI, Switzerland
Prof. MITTIG, Wolfgang	mittig@nscl.msu.edu	MSU-NSCL, USA
Mrs. MITU, ANDREEA	andreea.mitu@nipne.ro	IFIN-HH, Romania
Mr. MORRALL, Paul	paul.morrall@stfc.ac.uk	STFC, UK
Dr. MYHRE, Kristian	myhrekg@ornl.gov	ORNL, USA
Dr. PELLEMOINE, Frederique	pellemoi@frib.msu.edu	MSU-FRIB, USA
Dr. PENTTILA, Heikki	penttila@jyu.fi	University of Jyväskylä, Finland
Dr. SCHAUMANN, Gabriel	gschaumann@ikp.tu-darmstadt.de	Technische Universität Darmstadt
Dr. SCHUMANN, Dorothea	dorothea.schumann@psi.ch	PSI, Switzerland
Mrs. SIBBENS, Goedele	goedele.sibbens@ec.europa.eu	EC JRC-Geel, Belgium
Mr. STESKI, Dannie	steski@bnl.gov	BNL, USA
Dr. STODEL, Christelle	stodel@ganil.fr	GANIL, France
Dr. STOLARZ, Anna	anna@slcj.uw.edu.pl	HIL-UW, Poland

Ms. STONER, Constance	constance@acf-metals.com	Arizona Carbon Foil Co., Inc. ACF-Metals, USA
Dr. STONER, John	metalfoil777@gmail.com	Arizona Carbon Foil Co., Inc. ACF-Metals, USA
Mr. STONER, Robert	contact@acf-metals.com	Arizona Carbon Foil Co., Inc. ACF-Metals, USA
Dr. STRACENER, Daniel	stracenerdw@ornl.gov	ORNL, USA
Mr. VANLEEUW, David	david.vanleeuw@ec.europa.eu	European Commission JRC-Geel
Dr. VERMEULEN, Etienne	etienne@lanl.gov	EC JRC-Geel, Belgium
Mr. WETERINGS, Wilhelmus	wim.weterings@cern.ch	CERN, Switzerland
Dr. YANAGISAWA, Yoshiyuki	yanagisa@riken.jp	RIKEN, Japan
Dr. YOSHIDA, Koichi	yoshida@ribf.riken.jp	RIKEN, Japan
Dr. YOSHIMOTO, Masahiro	yoshimoto.masahiro@jaea.go.jp	JAEA / J-PARC center, Japan
Dr. ZACH, Mike	m4z@ornl.gov	ORNL, USA
Dr. ZEISLER, Stefan	zeisler@triumf.ca	TRIUMF, Canada

Statement of Peer review

In submitting conference proceedings to *Web of Conferences*, the editors of the proceedings certify to the Publisher that

1. They adhere to its **Policy on Publishing Integrity** in order to safeguard good scientific practice in publishing.
2. All articles have been subjected to peer review administered by the proceedings editors.
3. Reviews have been conducted by expert referees, who have been requested to provide unbiased and constructive comments aimed, whenever possible, at improving the work.
4. Proceedings editors have taken all reasonable steps to ensure the quality of the materials they publish and their decision to accept or reject a paper for publication has been based only on the merits of the work and the relevance to the journal.

Title, date and place of the conference

29th International Conference of the International Nuclear Target Development Society (INTDS2018)
8-12 October 2018
Michigan State University - Facility for Rare Isotope Beams, East Lansing, MI, USA

Proceedings editor(s):

Frederique Pellemoine
Wolfgang Mittig
Christelle Stodel

Date and editor's signature

20-Feb-2020

Progress status in fabrication of HBC stripper foil for 3-GeV RCS at J-PARC in Tokai site

Masahiro Yoshimoto*, *Yoshio* Yamazaki, *Takamitsu* Nakanoya, *Pranab Kumar* Saha, and *Michikazu* Kinsho

JAEA/J-PARC, Japan Atomic Energy Agency, 2-4 Shirane, Shirakata, Tokai, Naka-gun, Ibaraki, 319-1195, Japan

Abstract. In the 3-GeV Rapid Cycling Synchrotron (RCS) at Japan Proton Accelerator Research Complex (J-PARC), we adopted thick Hybrid type Boron-doped Carbon (HBC) stripper foil for multi-turn H^- charge-exchange injection. The HBC stripper foil developed at KEK has demonstrably improved the foil lifetime significantly. In addition, Graphite Thin Films (GTF) developed by KANEKA corp. was attempted to use for user operation. The GTF is a multilayer graphene sheet and it is expected to be a new candidate for highly durable charge stripping foil. HBC and GTF foils can be both employed the J-PARC RCS for long-term user operation. Early manufacturing of the stripper foil for the J-PARC was carried out according to the following two steps: foil fabrication at the KEK Tsukuba site and foil preparation at the JAEA Tokai site. Recently, the HBC foil deposition apparatus was relocated from KEK to JAEA, and new HBC foils were fabricated at the Tokai site. The performance of the new HBC foil was verified to be equivalent to the original HBC foil with offline beam irradiation tests. Furthermore, the use of the new HBC foils was successfully demonstrated during RCS user beam operation for one week.

1 Introduction

The Japan Proton Accelerator Research Complex (J-PARC) is a multipurpose high intensity proton accelerator facility [1, 2]. It is a joint project between Japan Atomic Energy Agency (JAEA) and High Energy Accelerator Research Organization (KEK), and it is located in Tokai-Mura, Japan. J-PARC consists of three accelerators, namely, a 400-MeV LINAC, a 3-GeV Rapid Cycling Synchrotron (RCS), and a 30-GeV Main Ring Synchrotron (MR), and three experimental facilities, namely, a Materials and Life Science Experimental Facility (MLF), a Hadron Experimental Hall (HD), and a Neutrino Beam Line to the Super-Kamiokande (NU). In this chain of accelerators, the RCS performs two functions; as a proton driver to produce pulsed muons and neutrons at MLF and as an injector to MR, aiming at 1 MW output beam power. To achieve MW class high power proton beams in at J-PARC, the multi-turn charge-exchange H^- beam injection scheme with stripper foils has been adopted for RCS.

In RCS, a thick Hybrid type Boron-doped Carbon (HBC) stripper foil has been in mainly use since the start of early beam commissioning in 2007 [3]. The HBC stripper foil was developed by Sugai at KEK to drastically improve foil lifetime [4]. The HBC foil technology is patented [5], where the original assignee is KEK, and the inventors are I. Sugai, M. Oyaizu, and H. Takeda. Then, JAEA entered into a licensing arrangement with KEK for the HBC foil in 2018. The HBC foil deposition apparatus was relocated from the KEK Tsukuba site to the JAEA Tokai site, and we have started fabrication of a new HBC foil. (We call the new foil fabricated at the JAEA Tokai site J-HBC foil.).

In this paper, we report the satisfactory results of the original HBC foil obtained over long-term user operation of RCS. In addition, the results of various offline tests for the performance evaluation of the new J-HBC foil and trial use in J-PARC RCS are reported.

2 Stripper foil in J-PARC RCS

The stripper foil plays a very important role in a high power proton beam accelerator, and it is one of the most important components in J-PARC. The HBC stripper foil, which is produced by the arc-discharge method with carbon electrodes doped with boron carbide (B_4C) particles, has adequate toughness to withstand high-intensity beam irradiation. Indeed, more than 10 sheets of HBC foils have been used in user beam operation, and all of them have withstood beam irradiation thus far. Therefore, we consider that the HBC foil has achieved satisfactory results in the J-PARC. By contrast, Graphite Thin Films (GTF) supplied by KANEKA Corp. [6, 7] have been trialled recently. GTF is a multilayer graphene sheet and it is prepared by annealing a thin polyimide film at temperatures of approximately 2900 °C in vacuum, and it is characterized by high tensile strength of approximately 40 MPa. The preparation process endows GTF with excellent mechanical robustness and flexibility, high purity, and superior heat resistance. Thus, GTF is expected to be a new candidate for highly durable charge stripping foils. Indeed, the GTF stripper foil is installed in the RCS and used for user operation.

* Masahiro Yoshimoto: yoshimoto.masahiro@jaea.go.jp

As the result, it has been verified to have high robustness against high power beam irradiation, but, at the same time, an issue which is discussed below has been found in the RCS.

The beam commissioning of the RCS for initial beam tuning started in October 2007, and user operation with an output power of 4 kW started in December 2008 [8]. Since then, the output beam power has been increasing smoothly with beam tuning to reduce beam losses. Two important milestones at J-PARC from the viewpoint of increasing the output beam power are the following [9]: the energy upgrade of LINAC from 181 MeV to 400 MeV during the long shutdown in 2013, and the peak current upgrade from 30 mA to 50 mA by renovating the front end devices in the LINAC during the summer shutdown in 2014. After the two important upgrades, continuing efforts to increase the output beam power while minimizing beam losses have been sustained. The first 1-MW beam trial test was performed successfully in October 2014, immediately after the injection beam current upgrade and output beam power for user operation was stood at up to 500 kW in 2015. After performing fine beam tuning tests to reduce beam losses, 1-MW beam operation at 25 Hz is now estimated to be realized sufficiently within a permissible beam loss level [10].

The HBC stripper foils have been supporting the RCS beam tuning and user operation over a period of 10 years. Figure 1 shows a summary of the actual accomplishments of HBC stripper foils in the RCS since recovery from the huge earthquake in 2011 until the summer shutdown in 2016. The upper horizontal bar chart shows the status of user operation. The middle two plots show as follows: the trend of the number of injected H^- particles into the RCS every hour, indicating the total output beam power of RCS beam operation; the trend of integral of the injected H^- particles irradiating onto each stripper foil. However, the actual magnitude of beam irradiation onto each foil is five to seventy times larger than integral value in the graph because not only the injected H^- beam but also the circulating proton beam irradiate onto the foil [10]. The lower photos show the changes in the stripper foils owing to beam irradiation. The stripper foils from ID (a) to ID (h) are the HBC foils, and ID (i) is the GTF foil. The stripper foils have been deformed by beam irradiation, but none of them got broken during beam irradiation thus far. In most cases, irradiated foils have been replaced with new ones during every summer shutdown as a precautionary measure to avoid the spread of broken fragments of highly activated stripper foils in the beam ducts.

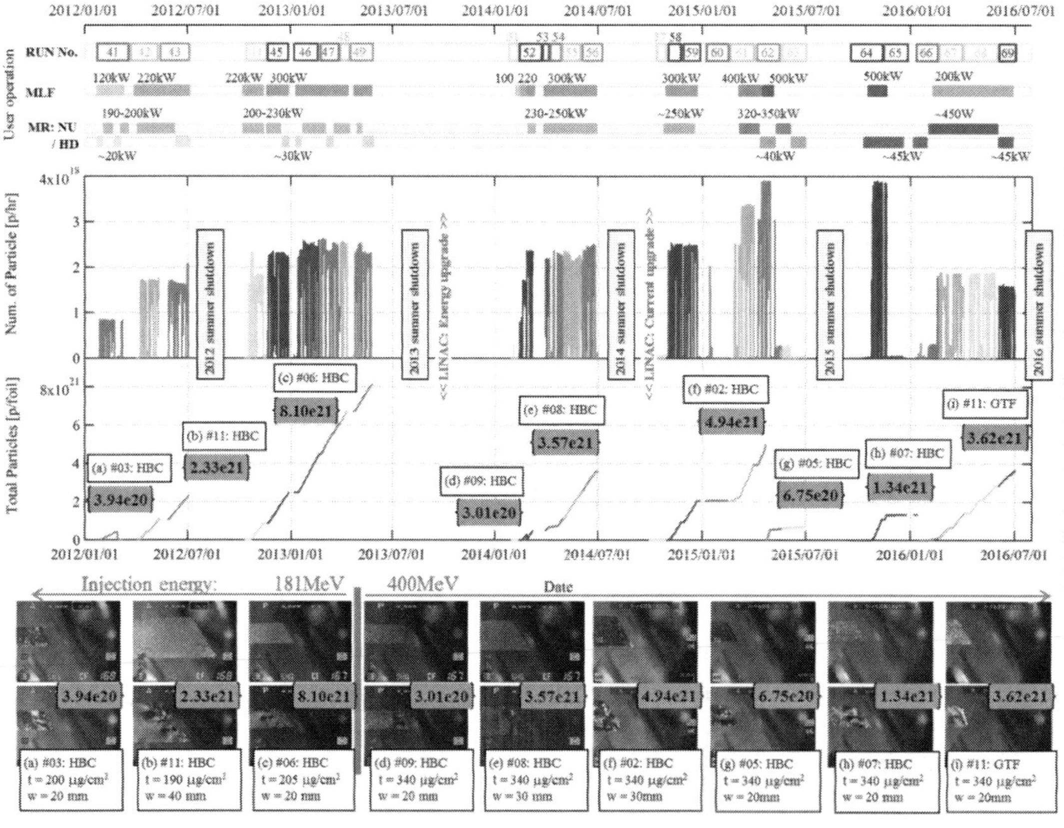

Fig. 1. History of beam power increase and stripper foil usage in the J-PARC RCS from 2011 to 2016.

In J-PARC RCS, the stripper foil is semi self-standing and is not restricted with the foil frame. Therefore, observed transformation due to beam irradiation shows interesting properties of the foil itself. The shape and the surface of the HBC foils show drastic changes. The irradiation spot is flattened, and the micro bump structure on the surface disappeared. By contrast, the surrounding areas were shrunken and wavy in appearance. It is assumed that the wave appearance can be scribed by pulling from the irradiation spot. However, these drastic changes have not affected the charge exchange property of the foils. On the other hand, the changes in the GTF stripper foil are different from those in the HBC foils. The GTF stripper foil was installed in the RCS in September 2016 and has been employed for user operation with an output beam power of 150 kW for 4 months. The beam irradiation spot shows almost no changes. The charge exchange efficiency shifted gradually for that period, but the changes does not contribute to stop the user operation. The foil was confirmed to be robust against high power beam irradiation. However, the area surrounding the beam irradiation spot is bent, and especially, the edge of the foil is curved significantly. The GTF foil is a multi-layer graphene sheet, and it has a very ordered structure. Therefore, effect of material damages, such as lattice defects caused by beam irradiation, cannot be contained locally, and they spread across the entire foil in the form of internal stresses. Curving of the foil edge is a critical issue for RCS because it leads to leak the injecting H^- particles without hitting the foil and to increased foil scattering beam losses. Thus, a foil lifetime for charge exchange beam injection should be determined by the foil curving. To evaluate the actual lifetime of the GTF stripper foil, we must continue our investigation of its use in user beam operation.

3 Fabrication of New HBC foil for J-PARC

3.1 Manufacturing process of HBC foils

Stable supply of the stripper foils is important from the viewpoint of sustaining stable beam operation in J-PARC. Early manufacturing process of the HBC stripper foil at J-PARC was carried out according to the following two steps, as shown in figure 2: foil fabrication at KEK Tsukuba site and foil preparation at JAEA Tokai site [11].

The foil fabrication step consists of three processes. For pre-processing, a glass plate is coated with a release agent called "Crème-Cote" [12] to release the deposited foil from the glass plate easily. The main processing is carbon evaporation by using the arc-discharge method. By using the boron doped carbon electrodes, HBC foils can be fabricated. After evaporation processing, the foils deposited on the glass is annealed at approximately 470 K for 2 h. This post-processing is performed to release the high residual mechanical stresses in the deposited foil. In addition, it is an important processing for reduction the curving of the foil due to beam irradiation.

The foil preparation step consists of five processes. For releasing the deposited foil, the glass plate is placed carefully in pure water to float the foil on the water. Therefore, the floating foil is collected from the water. By pulling up the center line of the foil, a double-layer structure can be created [13]. After drying, the collected foil is placed between two pieces of thin tracing paper and is cut into the desired size by using scissors. Finally, a pair of U-shaped aluminium frame is used, with bonded SiC fibers of 7 μm diameter, and the foil is mounted on the frame [14]. Fifteen stripper foils on the frames are set in the frame rack and installed in the vacuum chamber as spare foils during user operation.

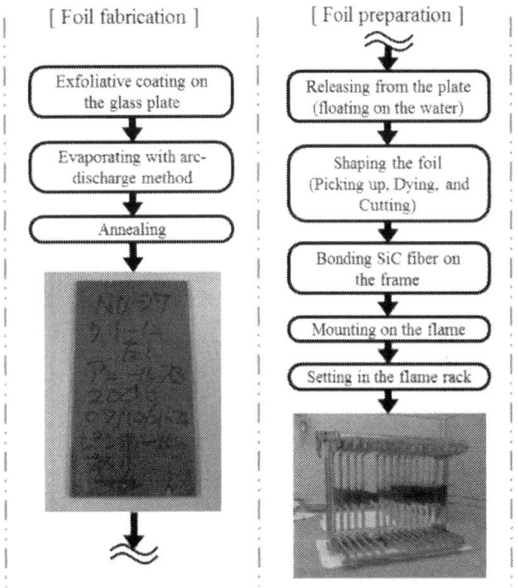

Fig. 2. Schematic diagram of foil manufacturing process.

3.2 Relocation and restoration of deposition apparatus for HBC foil

The foil fabrication step significantly influences the performance of the HBC foil. Therefore, it was executed recently at the KEK Tsukuba site by Sugai who is the inventor of and an expert on the HBC foils. However, it was difficult to fabricate the HBC foils at KEK. To maintain a stable supply of the HBC foils at J-PARC, there was a need to quickly restart foil fabrication at JAEA Tokai site. Before starting work on the HBC foil fabrication, JAEA entered into a licensing arrangement with KEK for manufacturing the HBC foils. Subsequently the device transfer procedure from KEK to JAEA was completed. Then, the deposition apparatus for the HBC foils was relocated from the Tsukuba site to the Tokai site.

This apparatus had been left unused without any maintenance for a considerable period since it was last used. Therefore, the equipment exhibited multiple problems, such as water leak, vacuum leak, metal corrosion, and dielectric breakdown. After relocation, machine maintenance, part replacement, and overhaul were performed. Devices with unnecessary functions were removed. As a consequence of many efforts, the deposition apparatus was restored successfully after one year interruption. Figure 3 shows the restored deposition apparatus after it was overhauled.

Some vapor deposition conditions for new foil fabrication are different from the original conditions because of repair and overhaul of the deposition apparatus. Furthermore, detailed procedure to fabricate the HBC foils prescribed by Sugai is not available. In such an environment, we started an attempt to fabricate the new foil at the Tokai site by recalling how he had fabricated the HBC foils previously. And the vacuum components were treated with extra care to prevent any contamination. For example, the glass plates were washed with an ultrasonic cleaning machine, and the deposited foils both before and after their release from the glass plate were stored in a desiccator. After several trial deposition tests, we successfully fabricated a new HBC foil at the JAEA Tokai site. (As mentioned before, we call the new foils J-HBC foil.) Subsequently, the performance of the J-HBC foil was evaluated before installation in the RCS and to use for the user operation.

Fig. 3. Photograph of restored deposition apparatus for HBC foil based on EBX-1000 system.

3.3 Performance evaluation of J-HBC foil

The original HBC foil was developed to improve foil lifetime against high power beam irradiation and it can achieve actual accomplishments at the long-term user operation in the J-PARC RCS. By contrast, we still cannot understand completely why the carbon foils can obtain higher durability against beam irradiation by boron-doping. To analyse the HBC foil by various methods and to understand the mechanism of foil breakage owing to beam irradiation, we have been conducting offline beam tests at Takasaki Ion Accelerators for Advanced Radiation Application (TIARA) in National Institutes for Quantum and Radiological Science and Technology (QST) [11, 14].

Therefore, same offline tests could be conducted for performance evaluation of the J-HBC foil by comparing the test results with those obtained using the original HBC foils. Hence, the following four offline tests were performed to evaluate the performance of the J-HBC foil: (1) surface microstructural observation by means of Scanning Electron Microscopy (SEM), (2) elemental composition analysis and foil thickness estimation by using the Rutherford Backscattering Spectrometry (RBS) method, (3) identification of trace impurities in the foil by using the micro-Particle Induced X-ray Emission (μPIXE) method, (4) endurance test and foil lifetime estimation by using argon ion beam irradiation and surface observation by digital microscope.

3.3.1 Surface microstructural observation by SEM

The HBC foil based on the cluster carbon foil was fabricated using the ACDC arc-discharge method [15]. Therefore, the HBC foil consists of amorphous carbon, and it has a cluster structure with mixed large clusters of 300 ± 200 nm derived from the cathode electrode and small clusters of 3 ± 2 nm derived from the anode electrode. Figure 4 shows a comparison of the SEM images (JEOL JSM-6700F) of the original-HBC foil and J-HBC. The cluster structure of the J-HBC foil was found to be reasonable compared to that of the original-HBC foil.

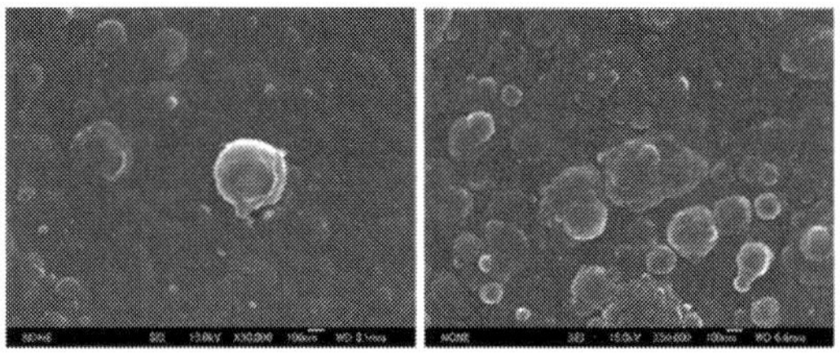

Fig. 4. Comparison of SEM images between original HBC foil and J-HBC foil. (left: original-HBC, right: J-HBC)

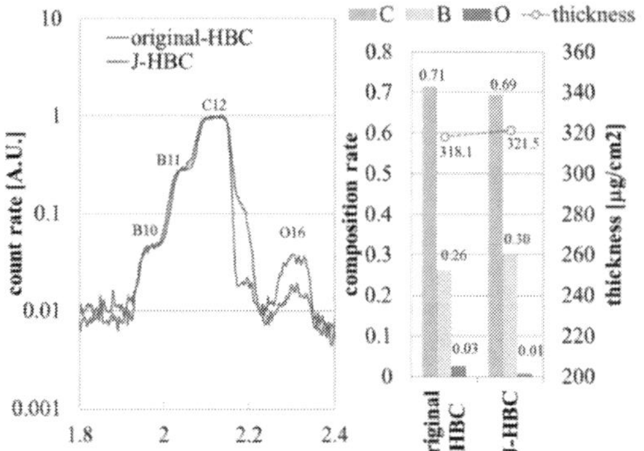

Fig. 5. Comparison of RBS spectra of original HBC foil and J-HBC foil (left), and analysis results for composition ratio and foil thickness estimation (right).

3.3.2 Elemental composition analysis and thickness estimation by RBS

The major elemental composition of the HBC foils was analysed using the RBS method with a proton beam of 3 MV in the TIARA 3-MV single-ended accelerator. The left graph in figure 5 shows a comparison of the RBS spectra of the original-HBC foil and the J-HBC foil. It is possible to obtain an analysis result for boron-to-carbon ratio and foil thickness estimation as shown the right bar chart of figure 5, by using SIMNRA backscattering spectra simulation code [16]. Both foils achieved the designed foil thickness of 330 ± 33 μg/cm^2. However, the boron-to-carbon ratios of the original-HBC and the J-HBC foils were slightly different. The cause of this difference is the electrode combination set in the deposition apparatus. In case of the original-HBC foil, the cathode electrode was a boron-doped (25 %) carbon rod of 10 mm in diameter while the anode electrode was a pure graphite rod of 15 mm in diameter. However, in case of the J-HBC foil, both electrodes were boron-doped (20 %) carbon rods of 10 mm in diameter.

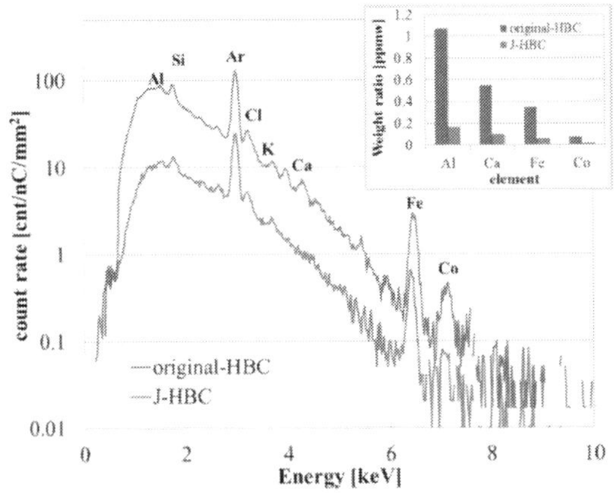

Fig. 6. Comparison of RBS spectra of original HBC foil and J-HBC foil (lower-left), and concentrations of trace impurities (upper-right).

3.3.3 Identification of trace impurities by μPIXE

Impurities in the foil, especially high-Z material, cause large-angle beam scattering and generate radionuclides in the foil. Thus, efforts to reduce impurities in the foil are very important from the viewpoint of using the foil in a high-power proton beam accelerator, such as J-PARC. To identify and evaluate quantitatively the trace impurities, the HBC foils were analysed with the μPIXE method. The lower-left graph in figure 6 shows a comparison of the μPIXE spectra of the original-HBC foil and the J-HBC foil. The same elements were identified in both foils. Argon, which exhibited relatively high peak signals, was identified as background noise by the measurement system. Other elements were estimated quantitatively as shown in the upper-right bar chart of figure 6. The concentrations of trace impurities in the J-HBC foil are considerably lower than those in the original-HBC foil, but the underlying reason is not clear. Both foils were fabricated according to almost the same manufacturing methods and using the same deposition apparatus except for the vacuum device treatment. These trace impurities are considered to mainly contaminate from the boron-doped electrode [11]. As a consequence, we will investigate the quantitative evaluation of the trace impurities depending on the arc-discharge condition.

3.3.4 Endurance test with ion beam irradiation

To investigate the endurance against beam irradiation, we performed an argon ion beam irradiation test by using the TIARA 400-kV ion implanter. In this test, the foil was irradiated with a 300-keV Ar^+ beam. It was found that the foil thickness of 330 μg/cm^2 was too thick for the Ar^+ beam to pass through. Then, a considerably thinner foil with a thickness of 20 μg/cm^2 was adopted to adjust an energy deposition in the foil to suit that at the J-PARC RCS. The thin J-HBC foil was mounted on an aluminium plate with a hole of 10 mm in diameter. After the J-HBC foil was irradiated for 30 min with a beam of 4 mm in diameter and beam current of 50 nA, its surface structure changed, but the foil did not break. Figure 7 shows the foil surface after beam irradiation, as observed with a digital microscope (KEYENCE VHX-6000). The irradiation spot was flattened and primary crack pattern on the surface disappeared. By contrast, the surrounding area showed radial wrinkles but the primary crack pattern is still present on the surface. This transition of the surface structure was the same as that observed in case of the original HBC foils [14]. It was analytically shown that radial wrinkles in a thin elastic sheet arise from compressive stresses in a region between inner and outer annular boundaries [17]. The radial wrinkles on the foil around the irradiation spot are ascribable to the same principle. Specifically, beam irradiation should cause an increase in density to the foil. And it brings a volumetric shrinkage and imposes inward radial force along the beam boundary.

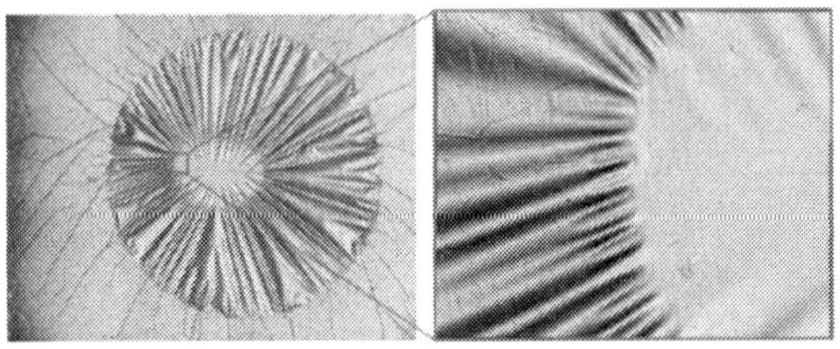

Fig. 7. Surface structure after Ar^+ beam irradiation, as observed using digital microscope.

3.4 Demonstration of J-HBC foil for RCS user operation

The J-HBC foils were verified to have good performance equivalent to that of the original HBC foil by means of offline tests in the TIARA. Thus, the J-HBC foils were installed in the RCS for user operation in September 2017. The RCS delivered protons to MLF users with a beam power of 300 kW and a repetition rate of 25 Hz, and to NU users with equivalent beam power of 700 kW and four shots at intervals of 2.48 s. Before use of the J-HBC foil for user operation, short duration trial tests were carried out under each beam condition. Figure 8 shows a photograph of the transition of the J-HBC foil during beam irradiation in the J-PARC RCS. Photo (a) shows the initial state of the foil. After the trial test with the MLF user beam for 30 minutes, the irradiation spot hollowed out slightly, as in photo (b). After the trial test with the NU user beam for 5 min, the foil deformed clearly, as in photo (c). However, its performance of the stripper foil remained unaffected. Then, the J-HBC foil was successfully used in the user operation for 1 week, as in photo (d). This great result is a big milestone for HBC foil manufacture in J-PARC. In the future, we will attempt to use the J-HBC foil for long-term operation starting October 2018 to evaluate its performance.

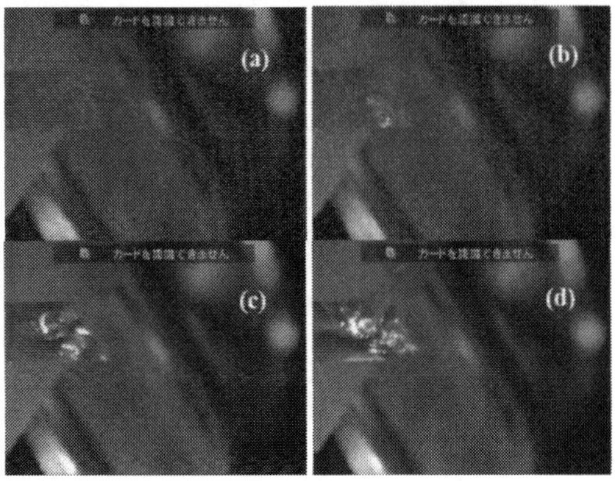

Fig. 8. Observation of J-HBC foil during user beam operation. (a) before irradiation, (b) trial test with MLF user beam for 30 min, (c) trial test with NU user beam for 5 min, and (d) after MLF and NU user operation for 1 week.

4 Summary

J-PARC RCS has been using the HBC stripper foils since the start of its operation, and the good performance of the stripper foil has been verified. Beam power for user beam operation has been increasing smoothly, and the HBC foil can be used for long-term user operation. The deposition apparatus for the HBC foil was relocated from the KEK Tsukuba site to the JAEA Tokai site in 2017 and we started fabrication of a new HBC foil at the JAEA Tokai site (J-HBC). Before using the J-HBC foil for user beam operation, offline beam test for the performance evaluation in the TIARA took place. The obtained test results are as follows: The cluster structure of the J-HBC foil is the same as that of the original HBC foil. The boron-to-carbon ratio of the J-HBC foil is slightly lower than that of the original HBC foil. The concentrations of trace impurities in the J-HBC foil are very low compared to those in the original HBC. Moreover, the endurance of the J-HBC foil against beam irradiation is the same as that of the original HBC foil. After the performance verification of the J-HBC foil, it was employed successfully in RCS user operation for 1 week. This great result is a major milestone for HBC foil manufacture at J-PARC. In the future, we will attempt to use the J-HBC foil for long-term operation to evaluate its performance. As a consequence, we have not only been carrying out stable supply for the J-PARC user operation but also done research and development on HBC foil to improve the performance and quality.

The authors would like to thank O. Takeda, R. Saeki, and M. Mutoh for technical assistance with foil manufacturing. The foil evaluation and analysis were mainly carried out in QST TIARA. We are grateful to Dr. T. Sato and M. Koka for special support for experiment and analysis of μPIXE, Dr. T. Taguchi for the analysis of the TEM and useful discussions, and Dr. S. Yamamoto for the analysis of the RBS and excellent support in TIARA. We are also grateful to Prof. Y. Irie for valuable advice from his rich experience and knowledge on accelerators including stripper foils. Finally, we would like to express our special thanks to Dr. I. Sugai for his guidance and encouragement. He pioneered the HBC stripper foil development and the success of this research is down to him.

References

1. High-intensity Proton Accelerator Project Team: JAERI Report No. JAERI-Tech 2003-044
2. High-intensity Proton Accelerator Project Team: KEK Report No. 2002-13.
3. M. Yoshimoto, P.K. Saha, Y. Yamazaki, M. Kawase, R. Saeki, N, Hayashi, K. Yamamoto, H. Hotchi, T. Ishiyama, M. Kinsho, I. Sugai, and Y. Irie, J. Phys.: Conf. Ser. **417**, 012073 (2013).
4. I. Sugai, Y. Takeda, M. Oyaizu, H. Kawakami, Y. Irie, Y. Arakida, K. Hara, H. Hattori, K. Kawasaki, J. Kamiya, M. Kinsho, Nucl. Instrum. Methods Phys. Res., Sect. A **561**,16-23 (2006).
5. I. Sugai, M. Oyaizu, and Y. Takeda, JP2007545272A, 2005-11-15. (in Japanese)
6. I. Sugai, M. Oyaizu, and Y. Takeda, WO2007/058224 A1, 2007-5-24. (in Japanese), (https://patents.google.com/patent/WO2007058224A1/en , accessed 2020-01-10, (in English))
7. A. Tatami, M. Tachibana, M. Murakami, T. Yagi, M. Murakami, AIP.: Conf. Proc. **1962**, 030005 (2018).
8. H. Hotchi, M. Kinsho, K. Hasegawa, N. Hayashi, Y. Hikichi, S. Hiroki, J. Kamiya, K. Kanazawa, M. Kawase, F. Noda,† M. Nomura, N. Ogiwara, R. Saeki, P. K. Saha, A. Schnase, Y. Shobuda, T. Shimada, K. Suganuma, H. Suzuki, H. Takahashi, T. Takayanagi, O. Takeda, F. Tamura, N. Tani, T. Togashi, T. Ueno, M. Watanabe, Y. Watanabe, K. Yamamoto, M. Yamamoto, Y. Yamazaki, H. Yoshikawa, and M. Yoshimoto, Phys. Rev. ST Accel. Beams **12**, 040402 (2009).
9. K. Hasegawa, H. Oguri, T. Ito, E. Chishiro, K. Hirano, T. Morishita, S. Shinozaki, H. Ao, K. Ohkoshi, Y. Kondo, J. Tamura, S. Yamazaki, T. Hori, F. Sato, Y. Nemoto, I. Koizumi, N. Ouchi, N. Kikuzawa, A. Ueno, A. Miura, S. Fukuta, A. Yoshii, K. Sato, A. Ozone, Y. Sawabe, Y. Kawane, H. Ikeda, Y. Ito, Y. Kato, K. Kikuchi, F. Hiroki, T. Takayasu, T. Usami, M. Yanai, K. Tadokoro, K. Ohsawa, F. Naito, Y. Liu, Z. Fang, T. Sugimura, K. Futatsukawa, K. Ikegami, M. Kawamura, K. Nanmo, Y. Fukui, T. Miyao, T. Maruta, A. Takagi, JPS Conf. Proc. **8**, 011012 (2015).

10. H. Hotchi, H. Harada, N. Hayashi, S. Kato, M. Kinsho, K. Okabe, P. K. Saha, Y. Shobuda, F. Tamura, N. Tani, Y. Watanabe, K. Yamamoto, M. Yamamoto, and M. Yoshimoto, Phys. Rev. ST Accel. Beams **20**, 060402 (2017).
11. Y. Yamazaki, M. Yoshimoto, P.Saha, M. Kinsho, T. Taguchi, S. Yamamoto, and I. Sugai, JRNC, **3**, 305, 859-864 (2015).
12. I. Sugai, M. Oyaizu, Y. Takeda, H. Kawakami, H. Hattori, and K. Kawasaki, Nucl. Instrum. Methods Phys. Res., Sect. A **590**, 32-36 (2008).
13. I. Sugai, et. al, Nucl. Instrum. Methods Phys. Res., Sect. A **613**, 457-461 (2010).
14. Y. Yamazaki, M Yoshimoto, O Takeda, M Kinsho, T Taguchi, S Yamamoto, T Kurihara, and I Sugai, J. Phys.: Conf. Ser. **417**, 012071 (2013).
15. I. Sugai, T. Hattori, H. Suzuki, H. Kinishita, H. Kato, K. Yamazaki, Nucl. Instrum. Methods Phys. Res., Sect. A **265**, 376 (1988).
16. M. Naterm SIMNRA User's Guide, Max Planck-Institute-fur-Plasmaphysik, ver.7.01 (2017).
17. B. Davidovitch, R. D. Schroll, D. Vella, M. Adda-Bedia, and E. Cerda, Proc. Natl. Acad. Sci. U.S.A. **108**, 18227 (2011).

Measurement of radioactivity and evaluation of activated nuclides due to secondary particles produced in stripper foil in J-PARC RCS

Masahiro Yoshimoto*, *Kota* Okabe, and *Michikazu* Kinsho

JAEA/J-PARC, Japan Atomic Energy Agency, 2-4 Shirane, Shirakata, Tokai, Naka-gun, Ibaraki, 319-1195, Japan

Abstract. Multi-turn charge-exchange beam injection is the key technique to achieve high-intensity proton beam accelerators. In J-PARC RCS, 400-MeV H^- beams from LINAC are injected into stripper foils such that most of the beams are converted into protons. The stripper foils are irradiated not only with the injected H^- beams but also with the protons circulating during the beam injection period. The high energy and intense beam irradiation into the foil generates secondary neutrons and protons via nuclear reactions. These secondary particles cause high residual activation around the stripper foils. Therefore, an activation analysis method that uses sample pieces is required to investigate the secondary particle. Furthermore, a new experimental system to measure secondary particles from the foil was installed at the L3BT 100BD line. Initial testing of the system has been completed, and beam testing will commence in October 2018.

1 Introduction

The 3-GeV Rapid Cycling Synchrotron (RCS) in Japan Proton Accelerator Research Complex (J-PARC) accelerates protons from a kinetic energy of 400 MeV to 3 GeV with a repetition rate of 25 Hz, and the aim is to achieve beam output power of 1 MW. The RCS performs two main functions: it serves as a proton driver for neutron/muon production at the Material and Life science experimental Facility and as a booster of the Main Ring synchrotron (MR) at the Hadron experimental facility (HD) and Neutrino experimental facility (NU) [1].

To achieve MW-class high-power proton beams in J-PARC, multi-turn charge-exchange H^- beam injection scheme with stripper foils has been adopted in RCS. Thick Hybrid-type Boron-doped Carbon (HBC) stripper foils have been mainly used since the start of beam commissioning in 2007 [2, 3]. The most important issue in achieving such MW-class high-power routine beam operation is maintaining machine activations within a permissible level, that is, preserving a maintenance environment without wearing special safety equipment to prevent workers from excess exposure. Therefore, we adopted the ring collimator system to remove the beam halo and to localize the beam loss in the collimator area [4]. Moreover, a large fraction of our effort has been dedicated to reducing and managing beam losses in J-PARC RCS [5]. To confirm the effect of beam loss localization, residual dose rates along the ring were measured periodically. The measurement results showed that unexpectedly high levels of radioactivity were found around the stripper foil [6].

By contrast, the following two important upgrades were applied to LINAC to achieve the desired output beam power of 1 MW from the RCS. The energy level was upgraded from 181 MeV to 400 MeV in 2013, and the peak current was upgraded from 30 mA to 50 mA in 2014 [7]. The first 1-MW beam trial test was performed successfully in 2014, right after the peak current upgrade. The output beam power for user operation reached up to 500 kW in 2015. After the two LINAC upgrades, finer beam tuning was performed to reduce beam losses in RCS. As a result, it was considered that 1-MW beam operation at 25 Hz can be realized with a permissible beam loss level [8]. However, radioactivity around the foil increased after the LINAC upgrade, and the residual dose rate around the foil could not be reduced drastically. Therefore, it is very important to understand the cause of this radioactivity and reduce the residual dose rate around the foil to achieve stable user operation in J-PARC.

2 Radioactivity in J-PARC RCS

2.1. Residual dose rate measurement

To investigate radioactivity in J-PARC RCS, the residual dose rate was measured frequently. Detailed dose rate distributions were obtained via dose rate measurement in contact with the beam ducts by using Geiger–Muller (GM) counters. These specific dose rate distributions clearly indicate the beam loss mechanisms in RCS.

* Masahiro Yoshimoto: yoshimoto.masahiro@jaea.go.jp

2.1.1 Residual dose rate distribution along ring

The J-PARC RCS has a three-fold symmetry lattice, and there are three straight sections (injection, extraction, and RF section) and three arc sections, as shown in Figure 1. Beam collimator systems, which can localize beam loss, are installed in the injection section. Figure 2 shows the typical measured residual dose rate distributions along the ring. The results indicate that the beam losses can be localized in the collimator successfully, and there are no high-level residual dose rates along the ring, except around the stripper foil. Therefore, the residua dose rate around the foil is not caused by standard beam loss in the ring, and the collimator system cannot reduce the dose rate any further. However, it is important to understand and reduce the residual dose rate around the foil for stable user beam operation in J-PARC.

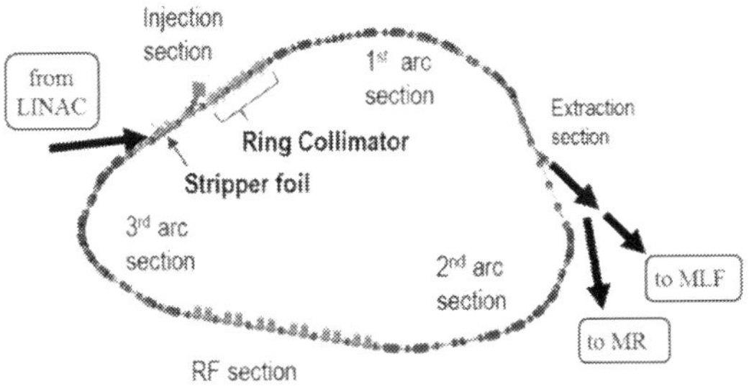

Fig. 1. Schematic view of J-PARC RCS ring.

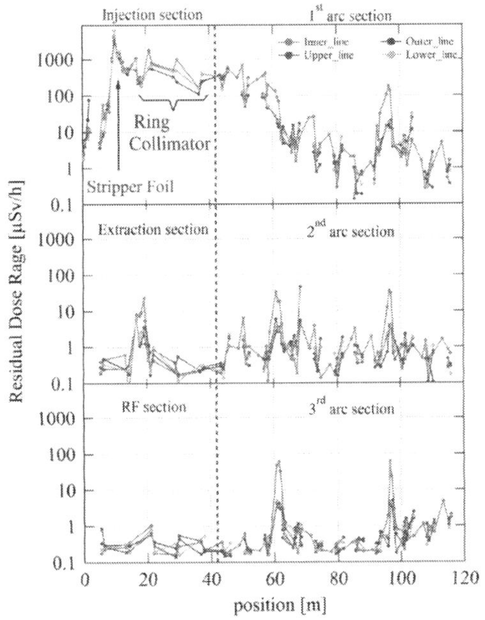

Fig. 2. Typical measurement results of residual dose rate distributions along the ring.

2.1.2 Residual dose distribution around stripper foil

Figure 3 shows a top view of the injection area in RCS. The beam orbits of the injected H^- beam and the circulating proton beams are merged by the bump magnets and irradiated onto the stripper foil during beam injection. Many beam injection components are installed in this small area.

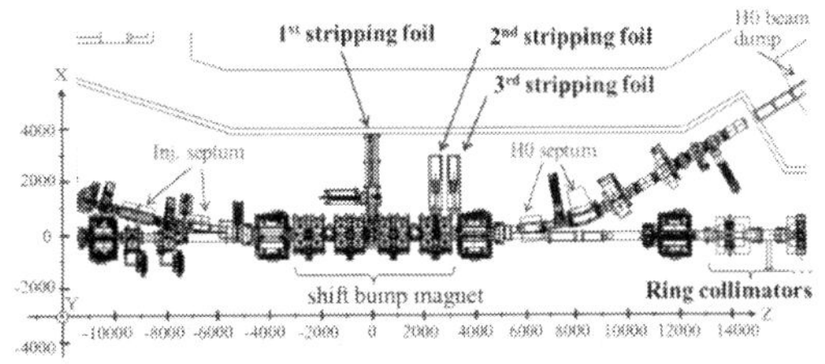

Fig. 3. Top view of injection area in RCS.

Figure 4 shows the typical measurement results of residual dose rate distributions around the stripper foil after the two upgrades to LINAC (energy upgrade and peak current upgrade). The upper graph shows a comparison between the mappings of the inner and the outer surfaces, and the lower graph shows a comparison between the mappings of the upper and the lower surfaces. The horizontal Z-axis represents the measurement position from the stripper foil. These mapping profiles have similar distributions measured at the injection energy of 181 MeV [6]. However, the dose rate magnitude is significantly higher.

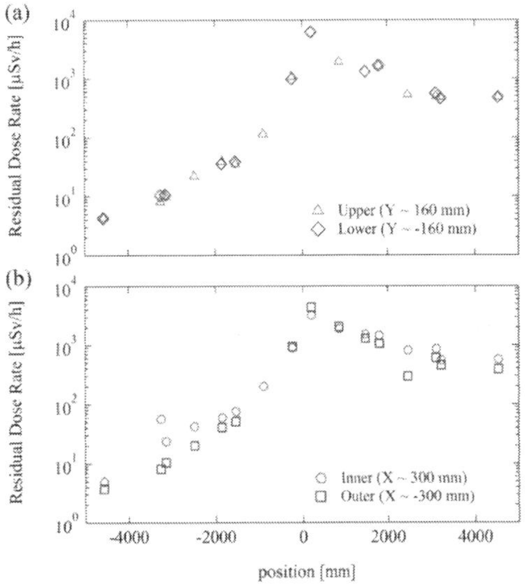

Fig. 4. Typical measurement results of residual dose rate distribution around stripper foil.

2.2 Nuclear reaction in stripper foil simulated using PHITS code

A simulation showed that the cause of the radioactivity around the foil for 181-MeV H^- beam injection was the presence of secondary protons and neutrons generated by nuclear reactions on the foil [6, 9]. Then, we simulated the interaction between the proton beam with injection energy of 400 MeV and the stripper foil by using PHITS Monte Carlo code [10]. PHITS is one of general purpose Monte Carlo particle transport simulation codes and has been used extensively. It can deal with the transport of all particles over wide energy ranges, using several nuclear reaction models, and nuclear data libraries. It has been previously compared with other Monte Carlo code (e.g., FLUKA) and confirmed its effectiveness [11]. Our simulation focused on the effects of particle scattering and nuclear reaction, and it did not consider the differences between H^- particles and protons. Thus, protons were set as the injected particles in the simulation. A mixture of 80% carbon and 20% boron was input as the foil material.

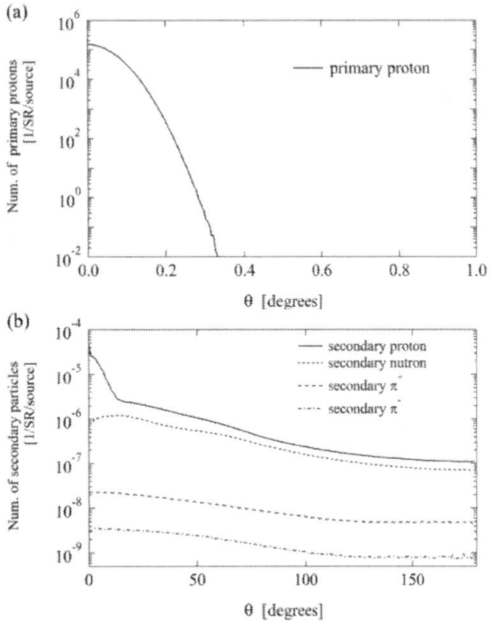

Fig. 5. Angle distribution of primary proton (a) and secondary particles (b) calculated using PHITS code.

Figure 5-(a) and (b) show the angle distributions of the primary proton and the secondary particles, respectively. Primary protons barely lost any energy, and almost all primary particles moved forward after passing through the foil, as in the simulation result obtained with the injection energy of 181 MeV. By contrast, the secondary particles emit from the foil to all directions. Therefore, they are capable of activating vacuum chambers and devices arranged nearest the stripper foil. After the energy upgrade from 181 MeV to 400 MeV, pions (π^+ and π^-) are newly generated. However, the number of secondary pions is considerably smaller than that of secondary protons and neutrons. Thus, the secondary protons and neutrons were focused on as the cause of the high-level radioactivity. Figure 6 shows the angle distributions of the secondary protons and neutrons with each categorized energy range. As these results, the dominant cause of radioactivity around the foil is assumed to be secondary protons and neutrons in the low-energy region.

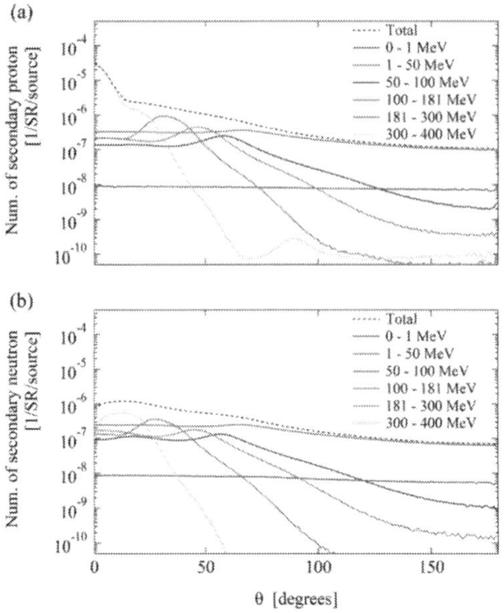

Fig. 6. Angle distribution of secondary proton (a) and secondary neutron (b) with such energy ranges calculated using PHITS code.

2.3. Radioactivity of foil and frame

2.3.1 Residual dose rate measurement of irradiation foil

In J-PARC RCS, irradiated and radioactive stripper foils can be collected from the ring and stored in a closed acrylic case for safety [6]. In addition, the stored foils can be subjected to various analyses.

Figure 7 shows residual dose rate mapping of the collected foils, as determined by comparing two sets of measurements performed using a GM counter (right after foil collection from the ring and one year after foil collection). The highest dose rate was observed at the beam irradiation spot on the stripper foil. By contrast, relatively high dose rates were measured in regions of the stripper foil and aluminum (Al) frame that were outside of the irradiation area. The physical aperture of the Al frame was considerably larger than the dynamic aperture, and thus, the injected H^- particles and circulating protons could not hit the frame. Then, two activation scenarios were assumed. In one of the scenarios, broken pieces of the activated foil were attached to the frame. In the other scenario, the secondary particles emitted from the foil activated the frame. In terms of reduction of radioactivity, the activation damping rate at the beam irradiation spot on the foil was higher than that at the frame. We assumed the presence of different radionuclides with different half-lives at the irradiation spot and the frame. Then, radioactivity analysis of the collected foil was performed to identify the radionuclides directly.

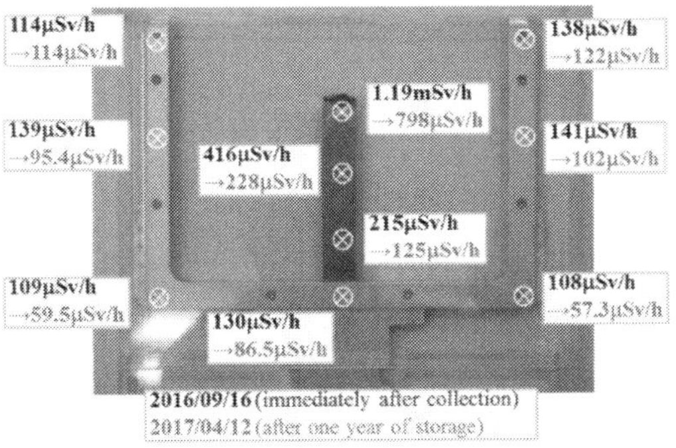

Fig. 7. Transition of residual dose rate measured on collected foil and frame.

2.3.2 *Identifying nuclides on stripper foil*

To identify the radionuclides on the collected stripper foil, activation analysis was performed using a portable germanium (Ge) semi-conductor detector (CANBERRA GR2018). To locate the measurement area that emitted gamma rays, the Ge detector head was attached the lead shielding collimator, and the lead shielding blocks were set in front of the foil, as shown in Figure 8. And, the gamma-ray spectra of the collected foil were obtained as shown in Figure 9.

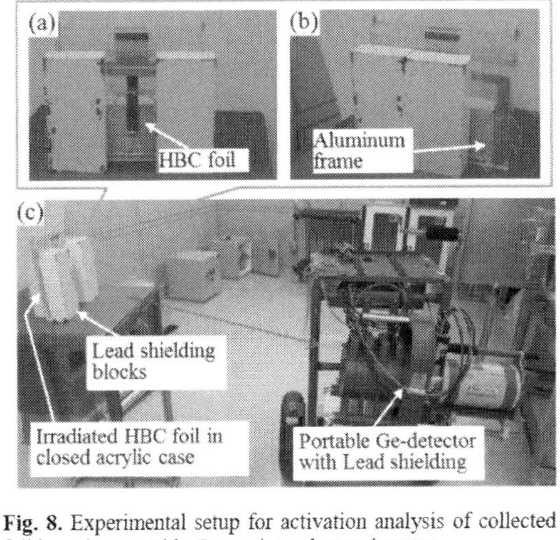

Fig. 8. Experimental setup for activation analysis of collected foil by using portable Ge semi-conductor detector.

The radionuclides in the beam irradiation area were mainly Be-7 and Na-22. Be-7 was produced from carbon or boron in the HBC foil. Na-22 in the area was produced from Al contamination in the foil or the 7 μm SiC fibers that supported the stripper foil [12]. By contrast, the only radionuclide on the frame was Na-22. This radionuclide was produced from the Al frame. If broken pieces of the activation foil were attached, Be-7 must be detected at the frame. Thus, the frame must have been activated by itself, and the cause of activation would have been secondary particles from the foil.

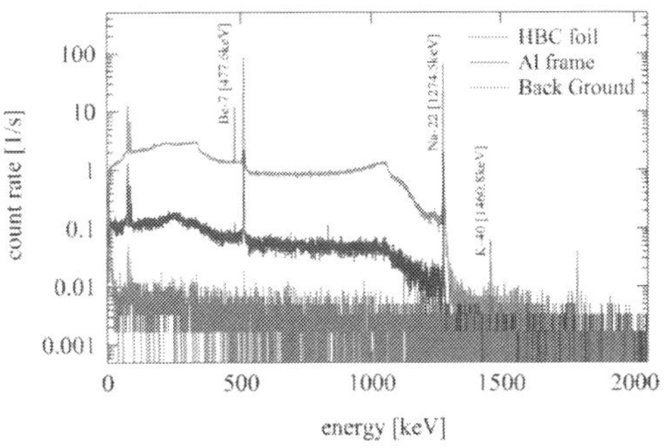

Fig. 9. Gamma-ray spectra in collected foil and frame.

3 Measurement of secondary particles on stripper foil

The highest radioactivity around the stripper foil was caused not by the standard beam losses but by the secondary particles generated in the stripper foil as a result of the nuclear reactions. Therefore, mitigating the beam losses may not be effective for reducing the radioactivity caused by this reason. In addition, the secondary particles must be measured experimentally, evaluated quantitatively, and managed effectively to preserve the maintenance environment for workers in this area.

3.1. Radioactivity analysis around stripper foil

The radioactivity analysis method is used commonly to determine trace concentrations of elements in samples irradiated with neutrons, charged particles, or photons [13]. In addition, it can be used to acquire information on the spatial distribution of a neutron field via neutron activation detectors [14]. We attempted to employ the radioactivity analysis method to identify high-energy particles. To investigate the secondary particles emitted from the foil, we applied the radioactivity analysis method to sample pieces obtained from around the foil. Pure copper plates measuring 20 mm in length, 5 mm in width, and 1 mm in thickness were prepared as secondary particle-irradiating targets. We performed measurements at four points, namely, upstream titanium (Ti) sleeve that connected the titanium flange and the ceramic beam duct for the shift bump magnet 1 (SB1u) [15], downstream sleeve for the shift bump magnet 2 (SB2d), upstream sleeve for the shift bump magnet 3 (SB3u), and downstream sleeve for the shift bump magnet 4 (SB4d). The angles between the beam direction and the direction of the four measured points seen from the stripper foil were approximately 175°, 135°, 50°, and 5°, respectively. The solid angles of the copper samples were approximately 1.00E-5 SR, 6.81E-4 SR, 7.34E-4 SR, and 1.03E-5 SR. The copper samples were set on the titanium sleeves. After user beam operation for one week, the sample pieces were collected and subjected to gamma-ray measurements by using a fixed-type Ge semi-conductor counter (ORTEC GEM25P4-70). Figure 10 shows the obtained gamma-ray spectra and nuclei histograms produced in the copper sample pieces. Various nuclei were observed, and the production rate of each nucleus as a function of the measurement point was diverse. If the production rate of each nucleus as a function of the energies of the irradiating particles is known, the energy and angle distributions of the secondary particles can be estimated.

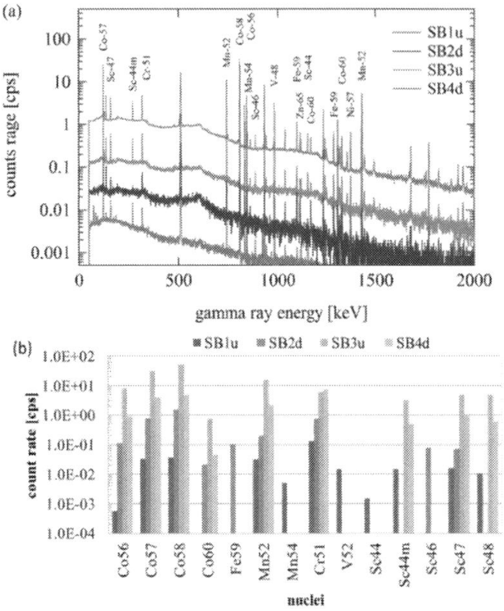

Fig. 10. Gamma-ray spectra and histograms of nuclei produced in copper sample pieces set on titanium sleeves of ceramics ducts.

3.2. Estimation of production rate of nuclei as a function of energy by using PHITS code

To investigate the energy and angle distributions of the secondary particles, the production rate of each nucleus as a function of the energies of the irradiating protons and neutrons energy must be known.

Experimental and theoretical results of production rates by proton- and neutron-induced reactions at intermediate energies (from a few MeV to a few thousand MeV) were previously published [16, 17]. In this work, however, we attempted to estimate the production rates by using PHITS code to easily compare with the measurement results in section 3.1. The energies of protons and neutrons irradiating the copper target were varied from 5 MeV to 400 MeV, and numbers of each type of nucleus produced were calculated. Figure 11 shows the histogram of each nucleus as a function of irradiation energy. There are clear differences in shape between proton and neutron irradiations.

By combining the simulation results of the production rate of nuclei with the experimental results of radioactivity analysis obtained using copper sample pieces, it is expected the energy and angle distributions of the secondary particles emitted from the foil can be evaluated. However, it is practically difficult to obtain accurate distributions for some reasons. The angle between the stripper foil and the irradiating H^- and proton beams change over the beam injection period. The secondary protons emitted from the foil cannot arrive at the copper samples directly because their orbits are bent by the shift bump magnet. When the copper samples are set on the outside of the beam ducts, undesired hadronic showers generated by secondary particles irradiating onto the beam duct affect the copper samples. Moreover, the production rate calculated using PHITS code has not been verified sufficiently in experimental settings.

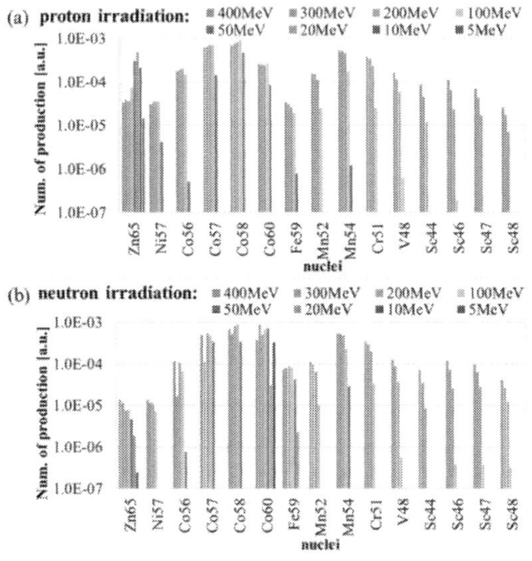

Fig. 11. Estimation of produced nuclei in sample pieces owing to proton and neutron irradiations.

3.3. New experimental system to measure secondary particles from foil

It is difficult to evaluate the secondary particles emitted from the foil by means of the radioactivity analysis method with sample pieces from around the foil. Similarly, it is difficult to measure the secondary particles directly in the injection area for following reasons: there is not enough space to set the required number of secondary particle detectors. Various beam losses occur in this area, and the radiation due to the beam losses disturbs the process of secondary particle detection. To solve the above issues, a new experimental site for various beam irradiation tests has been established at the L3BT 100deg-BD line. Moreover, a new secondary particle measurement system consisting of a foil-moving actuator and a detector vacuum chamber has been setup at the experimental site, as shown in Figure 12. The detector chamber houses multiple connection ports, and the sample pieces can be set inside the beam ducts. Then, secondary particle evaluation by using the radioactivity analysis method can be achieved via simple experiments. Simultaneously, we planned the measurement of secondary particles directly. A new particle detector system that can detect and classify protons, neutrons, electrons, and gamma rays, was developed. These experimental results are expected to be useful for verifying the accuracy of PHITS code.

Two methods to evaluate the secondary particles emitted by the foil are employed in our experiment. One is the direct measurement method to detect and identify each secondary particle, and the other is the radioactivity analysis method that uses several metal irradiation targets. In case of the first method, the development of radiation counters that can classify radiation as protons, neutrons, or gamma rays and analyse each energy level is vital. In case of the second method, it is important to select the target material and examine the radionuclides as indicators of secondary protons or neutrons in different energy ranges. Herein, we examine the radioactivity analysis method with PHITS code to search for radionuclides as indicators.

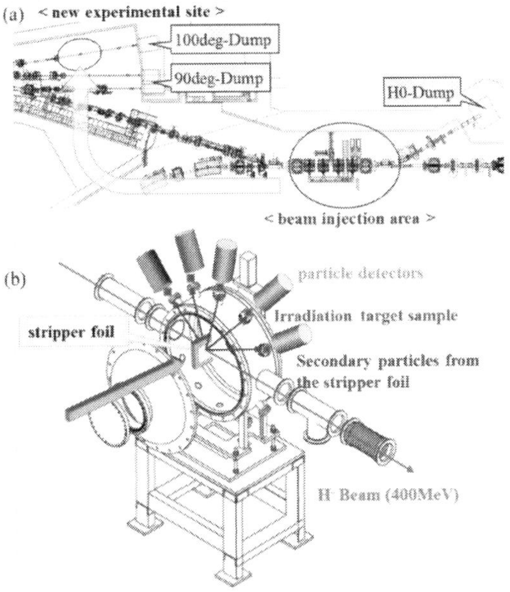

Fig. 12. Schematic views of experimental system to measure secondary particles from stripper foil. a) layout of experimental site, b) 3D-view of experimental device.

4 Summary

The stripper foil, which is the main element with respect to high-intensity proton beams, generates a large number of secondary particles, which lead to high levels of radioactivity. It is one of various important issues from the viewpoint of increasing the beam power of proton accelerators. To evaluate the secondary particles experimentally, detailed radioactivity analyses were performed around the foil in the injection area. However, it was difficult to perform in-situ measurements owing to various reasons. Thus, a new experimental system to measure the secondary particles from the foil was installed at the L3BT 100BD line. In experiments using this system, we plan to apply the radioactivity analysis method to sample pieces, in addition to measuring secondary particles directly by using the new particle detector. Initial testing of the system has been completed, and beam tests will commence from October 2018.

The authors would like to thank T. Nakanoya, O. Takeda, and R. Saeki for technical assistance with residual dose rate measurement and gamma-ray detection. We are grateful to T. Sugata and K. Nakamura from the University of Tokyo for special support for numerical analysis by using PHITS code. This work was supported by JSPS KAKENHI Grant Number JP16K05027.

References

1. High-intensity Proton Accelerator Project Team: JAERI Report No. JAERI-Tech 2003-044
2. M. Yoshimoto, P. K. Saha, Y. Yamazaki, M. Kawase, R. Saeki, N. Hayashi, K. Yamamoto, H. Hotchi, T. Ishiyama, M. Kinshoet, J. Phys.: Conf. Ser. **417**, 012073 (2013)
3. I. Sugai Y. Takeda, M. Oyaizu. H. Kawakami, Y. Irie, Y. Arakida, K. Hara, H. Hattori, K. Kawasaki, J. Kamiya, M. Kinsho, K. Kuramochi, Nucl. Instrum. Methods Phys. Res., Sect. A, **561**, 16-23 (2006)
4. K. Yamamoto, PRST-AB **11**, 123501 (2008)
5. H. Hotchi, H. Harada, N. Hayashi, M. Kinsho, P. K. Saha, Y. Shobuda, F. Tamura, K. Yamamoto, M. Yamamoto, M. Yoshimoto, and Y. Irie, PRST-AB **15**, 040402 (2012)
6. M. Yoshimoto, Y. Yamazaki, P. K. Saha, M. Kinsho, I. Sugai, Y. Irie, JRNC, **3**, 305, 865-873 (2015)
7. K. Hasegawa, H. Oguri, T. Ito, E. Chishiro, K. Hirano, T. Morishita, S. Shinozaki, H. Ao, K. Ohkoshi, Y. Kondo, J. Tamura, S. Yamazaki, T. Hori, F. Sato, Y. Nemoto, I. Koizumi, N. Ouchi, N. Kikuzawa, A. Ueno, A. Miura, S. Fukuta, A. Yoshii, K. Sato, A. Ozone, Y. Sawabe, Y. Kawane, H. Ikeda, Y. Ito, Y. Kato, K. Kikuchi, F. Hiroki, T. Takayasu, T. Usami, M. Yanai, K. Tadokoro, K. Ohsawa, F. Naito, Y. Liu, Z. Fang, T. Sugimura, K. Futatsukawa, K. Ikegami, M. Kawamura, K. Nanmo, Y. Fukui, T. Miyao, T. Maruta, A. Takagi, JPS Conf. Proc. **8**, 011012 (2015)
8. H. Hotchi, H. Harada, N. Hayashi, S. Kato, M. Kinsho, K. Okabe, P.K. Saha, Y. Shobuda, F. Tamura, N. Tani, Y. Watanabe, K. Yamamoto, M. Yamamoto, and M. Yoshimoto, Phys. Rev. ST Accel. Beams **20**, 060402 (2017)
9. E. Yamakawa M. Yoshimoto, M. Kinsho, JPS Conf. Proc. **8**, 012017 (2015)
10. T. Sato, K. Niita, N. Matsuda, S. Hashimoto, Y. Iwamoto, S. Noda, T. Ogawa, H. Iwase, H. Nakashima, T. Fukahori, K. Okumura, T. Kai,

S. Chiba, T. Furuta, L. Sihver, J. Nucl. Sci. Technol. **50**, 9, 913-923 (2013)

11. K. Kurosu, I. J. Das, V. P. Moskvin, Nucl. Inst. and Meth. B **367**, 14-25 (2016)
12. Y. Yamazaki, Y. Yamazaki, M. Yoshimoto, P. Saha, M. Kinsho, T. Taguchi, S. Yamamoto, I. Sugai, JRNC, **3**, 305, 859-864 (2015)
13. M. D. Glascock, *Instrumental Multi-Element Chemical Analysis,* 93-150 (1998)
14. M. Majerle, J. Adam, P. Caloun, S. A. Gustov, V. Henzl, D. Henzlová, V. G. Kalinnikov, M. I. Krivopustov, A. Krása, F Krízek, J. Phys.: Conf. Ser. **41**, 331-339 (2006)
15. M. Kinsho, D. Nishizawa, Y. Saito, H. Suzuki, H. Yokomizo, J. Vac. Sci. Technol. A **20**, 3, 829-831 (2002)
16. S. A M. Issa, M. A. M. Uosif, R. Michel, U. Herpers, P. Malmborg, B. Holmqvist, Nucl. Inst. and Meth. B **298**, 19-32 (2013)
17. I. Leya, R. Michel, Nucl. Inst. and Meth. B **269**, 2487-2503 (2011)

Measurements with the stripping foil test stand in the Linac4 Transfer Line

Chiara Bracco[1], *Louise* Jorat[1], *Remy* Noulibos[1], *Pieter* van Trappen[1], and *Wim* Weterings[1,*]

[1]CERN, CH-1211, Geneva 23, Switzerland

Abstract. In 2020, after the CERN accelerators complex Long Shutdown 2 (LS2), a novel Linac4 (L4)-to-PS Booster (PSB) charge-exchange injection system will allow to transform the L4 160 MeV H^- beam into H^+, which will be injected into the four PSB superposed rings. For this, a 200 μg/cm^2 carbon stripping foil will convert negative hydrogen ions (H^-) into protons by stripping off the electrons. L4 is now performing operational reliability runs, which include a stripping foil test stand installed in the L4 transfer line. These tests will permit to gain experience on the fragile foils, test different foil materials and thicknesses, measure the efficiency and lifetime of the foils, and evaluate the foil changing mechanism as well as the interlocking functions. This paper briefly describes the stripping foil test stand setup, before reporting on the obtained important test results.

1 Introduction

At CERN, the European Organization for Nuclear Research, particle accelerators and detectors are used to study the basic constituents of matter. Since its foundation in 1954, physicists and engineers have unified effort to increasing the discovery potential and pushing the limits of the energy frontier. The flagship accelerator of CERN is the Large Hadron Collider (LHC), world famous for being the most powerful particle accelerator ever built.

A massive improvement program of the LHC injector chain is presently being conducted under the LHC Injectors Upgrade (LIU) project [1] with the aim of producing the challenging High Luminosity LHC (HL-LHC) beam parameters [2]. The project comprises a new Linac, so-called Linac4 (L4), as well as major upgrades and consolidation of the Proton Synchrotron Booster (PSB), the Proton Synchrotron (PS) and the Super Proton Synchrotron (SPS).

L4 is a linear accelerator [3] intended to deliver a beam at 160 MeV energy to the four superposed synchrotron rings of the PSB. The main improvements of L4 stem from the use of negative hydrogen ions (H^-) instead of protons (H^+) with a higher injection energy than the currently used 50 MeV of Linac2 and the PSB will be equipped with an H^- charge-exchange injection system during the Long Shutdown 2019-2020 (LS2). Following the upgrade, H^- will be progressively injected horizontally into the PSB and converted into H^+ by passing through a 200 μg/cm^2 carbon foil to strip them of the electrons, aiming to convert ~98% of the H^- beam to H^+ [4]. Partially stripped H^0 and ~1% H^- missing the foil will be directed to an internal H^0/H^- dump located inside an injection chicane magnet [5]. This charge-exchange injection technique can achieve higher particle density providing an extremely flexible way to load particles into the PSB, making the accumulation of many turns possible with a tight control of the beam density [6].

In 2016 beam commissioning of L4 took place in steps of increasing energy to reach 160 MeV [7] and L4 is now performing operational reliability runs. A stripping foil test stand is installed in the L4 transfer line including the identical foil handling and exchange mechanism that will be used in the future PSB H^- charge-exchange injection system [8]. This will allow to gain experience using these fragile foils, test different foil materials and thicknesses, evaluate the lifetime of the foils and foil holders and test the foil changing mechanism as well as the interlocking functions prior to the final installation in the PSB in 2019.

*Corresponding author: wim.weterings@cern.ch

2 Stripping foil test stand

The design of the stripping foil handling and exchange mechanism, so-called TKSTR, of which an illustration is given in Figure 1, was extensively presented in [8, 9]. Based on commercially available components [10], the unit for the L4 test stand is identical as the mechanism intended to be installed in the PSB. It consists of a stainless steel belt, rotating over two pulleys, to which a maximum of six foil holders can be attached by use of quick disconnect sliders. This allows moving a foil into the beam aperture, with a perpetual rotation, so that each of the six foils can be re-selected into the nominal beam position with a precision of ±0.1 mm from which a foil movement in the horizontal plane of ±2 mm is possible in order to find the optimum position for operation. To this extend the mechanism is equipped with UHV compatible microswitches and membrane potentiometers, a detailed description of the electronics and controls for the TKSTR was presented in [11].

https://doi.org/10.1051/epjconf/202022901003

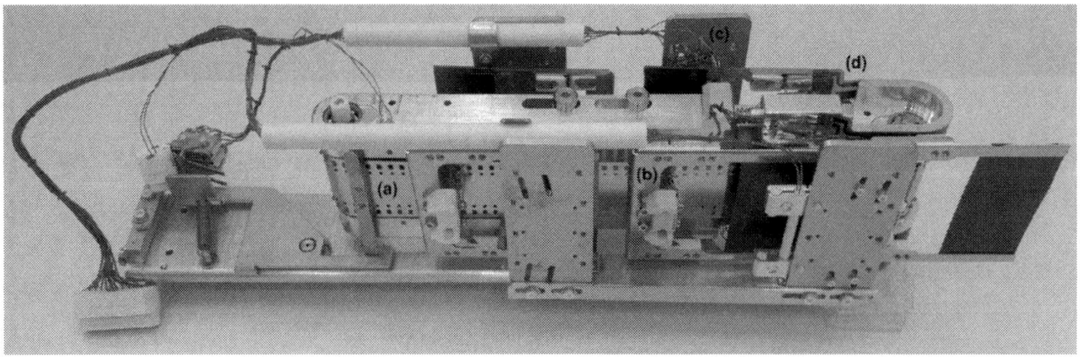

Fig. 1. The stripping foil exchange mechanism showing the rotating stainless steel belt (a) [10], the holders with stripping foils attached (b), the UHV compatible microswitches (c) and the membrane potentiometers (d) for the electronics and controls [11].

3 Test conditions

3.1 Beam characteristics

According to the latest specifications, L4 is supposed to reach a peak current of 40 mA. During nominal operation, some bunches will have to be chopped away to avoid losses during the transfer to the four superposed rings and during the longitudinal capture in the PSB. Finally, a peak current of 26 mA is expected at the PSB after chopping. Four pulses, one per ring, of up to 150 μs, will have to be produced with a flatness in current of ±5%.

All the foils at the test stand were tested with 160 MeV H^- beam with a transverse r.m.s. emittance of 0.4 π·μm·mrad mainly defined by the acceptance of the RFQ. Presently, a current of 30-35 mA has been reached in the linac and up to 20 mA could be transported to the foil. During the first dry run, in Spring 2018, the current at the foil was slowly increased from 5 mA to 20 mA and the pulse length from 200 μs to 600 μs. Initially, large current variations were observed along the pulse, when operating with 600 μs. Several optimizations were applied to the linac and the transfer line and a ±5% flatness could finally be achieved over the full pulse length. The quality of the stripping efficiency measurements over the Spring and Autumn Dry Runs reflect these more stable conditions and a clear reduction in the measured standard deviation was found.

In normal operation, in the PSB, each foil will see only pulses of maximum 150 μs but multiple crossings have to be considered, at slightly different locations of the foil due to the transverse painting, for the circulating beam. The beam conditions used during the tests, as summarized in Table1, allow to evaluate both the stripping efficiency and extrapolate the lifetime of each foil type during standard operation.

3.2 Stripping foils

The stripping foils that were used for the beam tests are commercially available and the characteristics are shown in Table 2. The required foil thickness for PSB injection is 200 μg/cm^2 (~1 μm) to ensure a theoretical stripping efficiency >99% while keeping the emittance increase below ~0.1 mm mrad for the LHC beam (required ~2 mm mrad rms normalised emittance at injection) and the uncontrolled beam losses below the 10^{-4} level [12].

The 32 mm wide foils are glued, on identical 58 mm vertical beam aperture stainless steel frames, with a mixture of 50% demineralised water and 50% Aquadag® 18% solution, of which a detailed description is presented in [9]. An illustration of the stripping foil holder attached to the rotating belt is given in Figure 2.

Table 1. Characteristics of the beam used for the tests.

Parameters	Value	Unit
Ion Species	H^-	
Energy	160	MeV
Repetition Rate	1.2	s
Pulse Length*	200-600	μs
Mean pulse Current	5-20	mA
Transverse emittance	0.4	π·μm·mrad
H/V beam size @ foil	~2.3/2.9	mm (radius)

*200μs to 600μs pulses with 50μs to 150μs batches

3.3 Measurements with BTC

Two beam current transformers (BCT), located upstream (L4T) and downstream (L4Z) of the foil, are used to measure the stripping efficiency by looking at their relative current decrease. Following the issues of a 20% higher signal at the downstream monitor, reported in [9], a cross-calibration of these BCTs at the percent level was done. The observed preamplifier saturation effects, due to a strong 352 MHz RF component well outside the presumed passband of the transformer, was corrected by the addition of appropriate low pass filters to have a reliable evaluation of the number of protons with respect to the original injected H^- [18]. Figure 3 shows a screen shot of the BCT control interface for operation and acquisition. The BCT positions L4T and L4Z respectively are the upstream and downstream BCTs with respect to the TKSTR.

 https://doi.org/10.1051/epjconf/202022901003

Table 2. Characteristics of the different foil types used.

Description	Weight	Remark	Name	
Arc evaporated amorphous carbon	200 $\mu g/cm^2$	Collodion coated	XCF-200	[13]
Arc evaporated amorphous carbon	199 $\mu g/cm^2$		GSI-199	[14]
Diamond-like carbon	200 $\mu g/cm^2$	Boron doped 10%	DLC-23-1000-S	[15]
Multilayer graphene	200 $\mu g/cm^2$		MLG-200	[16, 17]
Multilayer graphene	233 $\mu g/cm^2$		MLG-233	[16, 17]
Multilayer graphene	251 $\mu g/cm^2$		MLG-251	[16, 17]

Fig. 2. A MLG-251 stripping foil mounted on the stainless steel foil holder, attached to the rotating belt by use of a quick disconnect slider. The foil is 32 mm wide and the vertical beam aperture of the holder is 58 mm.

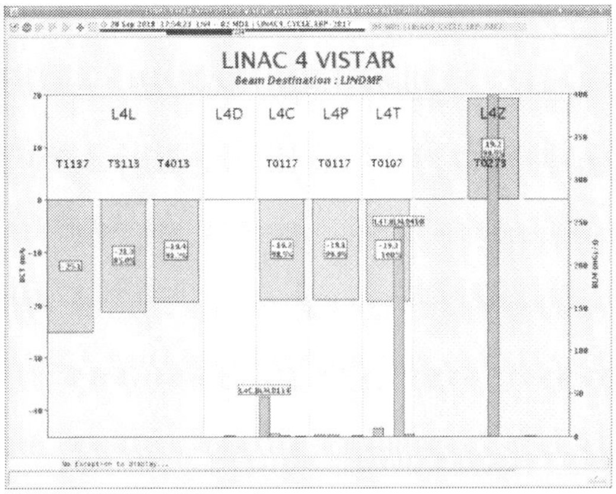

Fig. 3. Screen shot of the BCT control interface for operation and acquisition showing the beam current at different BCT positions in the L4. Location L4T (-19.3mA) and L4Z (19.2mA) respectively are the upstream and downstream BCTs with respect to the TKSTR. The readings of the Beam Loss Monitors (BLMs) are also shown.

A focusing quadrupole (L4T.RQF.021) for the H^- ions optics is installed after the stripping foil test stand. During the stripping foil tests, this quadrupole has to be switched off to maximize the transport of the protons when the foil is intercepting the beam and the status of charge is inverted. A clear effect on the measured stripping efficiency could be seen when the quadrupole was accidentally left on (see Figure 5).

3.4 Setting up with BTV

Correct position of the foil, with respect to the beam, is important for optimizing the stripping efficiency and for this reason a retractable optical beam observation system (BTV), consisting of a 1 mm thick Chromox (Al_2O_3 doped with CrO_2) scintillating screen, can be placed 6 mm in front of the foil [19]. The image of the screen, or of the stripping foil, is recorded by a radiation-hard camera, allowing checking either the beam position or the integrity of the foil [9]. Before measuring the stripping efficiency of a foil, a single beam pulse is shot on the BTV screen, to determine its position. Subsequently the foil edge can be moved ± 2 mm, or the beam optics adjusted, in order to achieve the theoretical stripping point position of the beam, which is at 14 mm from the edge of the foil.

During the first beam tests, that took place in October 2016, two foils (XCF-200) got broken while continues beam pulses were sent on the BTV during ~1 minute periods. After removing the screen, the foil reappeared damaged [9]. The mechanism that caused this foil disruption was thought to be the beam induced charging of the screen, and probably also the foil, during beam impacts causing rupture when the screen is moved. Since the improvement of the electrical contacts of the foil frames and the application of a silver conducting layer on the backside of the BTV screen this phenomena has not anymore been observed.

Another catastrophic failure occurred in September 2018. While setting up for tests with beam the BTV screen horizontally broke in the middle, see Figure 4. It is possible that continues beam shots on the screen, done to investigate the influence of the improved electrical contact and coating of the screen described above, have stressed the screen causing breakage. During a subsequent technical stop, inspection through the viewport showed the top part of the screen broken in two part at the bottom of the vacuum tank.

Fig. 4. Image from the radiation-hard camera of the broken BTV screen. The foil holder, ready to be put into beam position, can be seen at the right behind the screen. The glow on the top edge, right from the centre of the screen, is the position of beam impact previous to breakage.

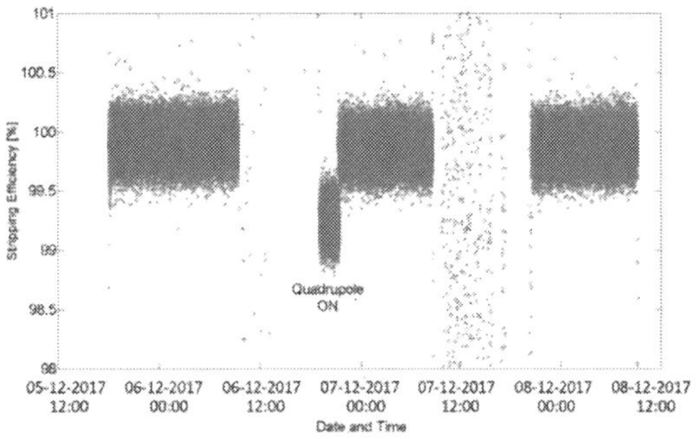

Fig. 5. Efficiency measurement data, from 3 night-shift runs, with 400μs pulses of ~15.7 mA beam current. The influence of incorrect beam optics, L4T.RQF.021 quadrupole on, can clearly be identified.

4 Test results

4.1 Stripping efficiency measurements

Throughout the L4 reliability run, stripping foil efficiency measurements took place during night-shifts and weekends, the day shift being dedicated to machine development and beam diagnostics. During the measurements, all L4T and L4Z BCT data was logged and post-processed to determine the stripping efficiency, Figure 5 shows the collected data of a typical run.

Taking into account data from several runs, and discarding all measurements outside the 98-100% data range, the stripping efficiency of the foils mentioned in Table 2 was calculated and is presented in Figure 6. One can observe that the efficiency of all foils is well above the theoretical value of >99%, as described in section 3.2, with the exception of MLG-200. According to experts, the observed behavior in terms of stripping efficiency and losses is as expected for multi-layer graphene foils [16]. Despite

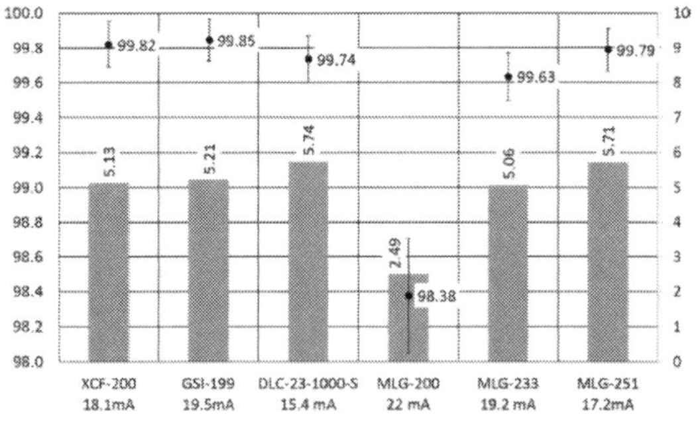

Fig. 6. Stripping efficiency measurements of the foils in the L4 test stand with 600μs beam pulses and indicated beam current. The left axis shows the efficiency [%] of the foils and the right axis the beam losses [mGy/s].

the same thickness of 200 μg/cm^2, as the other carbon-based foils, the crystalline structure of the graphene reduces the probability of interaction with the H^- ions and thus the charge-exchange process. On the other hand, the reduction of the cross sections for nuclear interaction, single and multi-Coulomb scattering determines the favorable effect of decreasing both the losses and the emittance blow up. For all these reasons, and for the interesting mechanical properties which make these foils relatively easy to handle, it was decided to test thicker graphene foils (MLG-233 and MLG-251) and assess their stripping efficiency. Unfortunately the emittance blowup induced by the different foils cannot be measured with the present installation since the effect of a single passage across the foil is below the resolution of the available diagnostics. Still the characterization of the foils in terms of losses and efficiency will provide useful information for bench-marking with simulations which will then allow to predict the expected effect on the final emittance in the PSB.

4.2 Foil damage and observations

Movement at the beam spot location of the DLC-23-1000-S and XCF-200 foils is observed during every beam shot, very well distinguished at the frequency of the L4 beam pulse. Inspection of the tested foils during a technical stop showed clear deformation and the creation of a crater, as shown in Figure 7, at impact location of foil DLC-23-1000-S. Nevertheless, subsequent tests showed that this phenomena has no influence on the stripping efficiency of the foil.

During tests, taking place in March 2018, it was observed that a DLC-23-1000-S foil, which had not been in beam position, was completely removed from the foil holder. Following inspection during the next technical stop, it was found lying inside the vacuum tank and looked completely burnt. Also delamination of foil layers was noticed at the edge of the foil, as shown in Figure 8, and the foil holder of this foil was particularly activated, > 38 μSv at contact, compared to 0.1 μSv of all other frames. High losses at the stripping foil test stand and a very low transmission to the downstream BCT were recorded, on March

Fig. 7. Appearance of a crater like deformation on a DLC-23-1000-S foil. Movement at the beam spot location is observed during every beam shot.

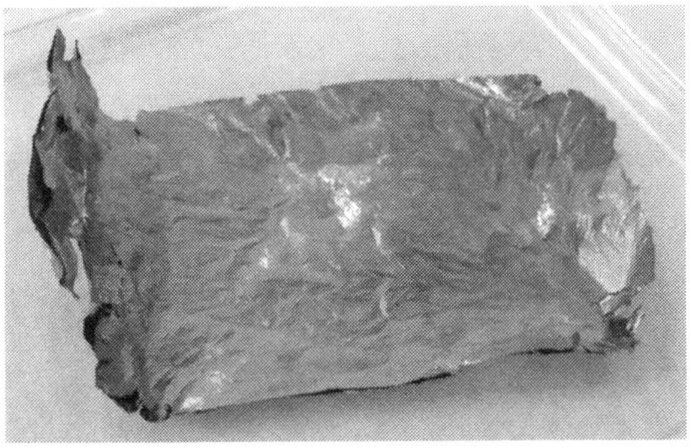

Fig. 8. DLC-23-1000-S foil that was detached from the foil without being tested with beam. The hypotheses is that the frame of this foil was hit by the showers of particles produced by mis-steered beam hitting the machine aperture, just upstream of the loader, during low energy tuning of the linac.

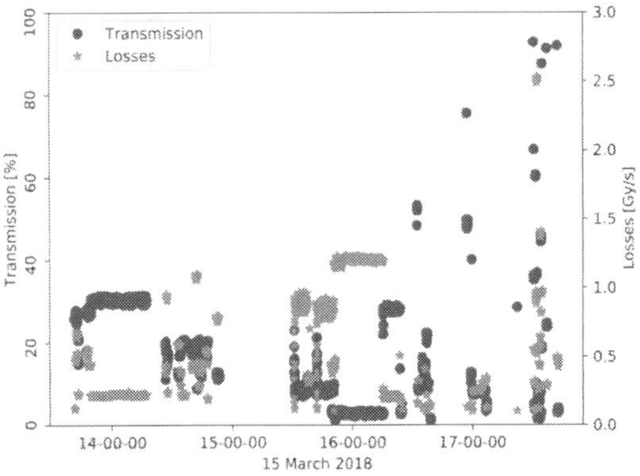

Fig. 9. Losses and current transmission during 3 MeV low energy tuning of the linac on March 15th. The ratio between the current of the upstream and downstream BCT values are shown in blue and the losses at the BLM located just in front of the stripping foil unit are shown in red.

Fig. 10. GSI-199 foil before (left) and after (right) one 600μs pulse of ~19.5mA. This effect had no influence on the stripping efficiency of the foil.

15th, during the low energy tuning of the linac cavities at 3 MeV as shown in Figure 9. Neither the foil or the BTV were intercepting the beam at that moment so the most likely source of these losses was a mis-steered beam hitting the machine aperture upstream of the test stand. The frame holding DLC-23-1000-S foil was facing the beam direction when the losses occurred and it might thus be fully hit by the electromagnetic secondary showers. This hypotheses was confirmed by damage that occured, due to irradiation, to the downstream electronics.

In October 2018, the GSI-199 type was tested for the first time. The foil was moved in beam position and after only one beam pulse the aspect of the foil completely changed, as shown in Figure 10, it looked like the foil shrunk or deformed towards the center of beam impact. Nevertheless, this had no influence on the performance of the foil, it remained in beam for 9 hours with a stripping efficiency of ~99.8%.

At another occasion, a foil frame got stuck behind a microswitch, but the system could be unblocked by reversing the movement of the stainless steel band. Nevertheless, this manoeuvre caused rupture of the DLC-23-1000-S foil.

5 Conclusion

A stripping foil test stand is installed in the L4 transfer line, including the identical foil handling and exchange mechanism that will be used in the new H^- charge exchange injection system of the PSB when L4 is connected with 160 MeV injection energy. During the L4 operational reliability runs the behaviour under beam conditions of several types of foils has been evaluated with measurement of their stripping efficiency using calibrated BCTs. All tested foil types have the expected theoretical stripping efficiency >99%.

Movement at the beam spot on foils is observed during every beam shot, very well distinguished at the frequency of the L4 beam pulse, and some foils show clear deformation at the impact location. This phenomena has no influence on the stripping efficiency of the foils.

Damage and increased radiation of a foil, which had not been in beam position, was observed and this phenomenon is thought to be caused by uncontrolled losses during the low energy tuning at 3MeV of the L4 beam impacting on the stripping foil mechanism.

The authors would like to acknowledge the help and support of the CERN Control Center (CCC) operators during the stripping foil efficiency tests.

References

1. E. Shaposhnikova et al., *LHC Injectors Upgrade (LIU) Project at CERN*, in *Proceedings, 7th International Particle Accelerator Conference (IPAC 2016): Busan, Korea, May 8-13* (2016), p. MOPOY059
2. G. Apollinari, O. Brüning, L. Rossi, *High Luminosity LHC Project Description*, in *Report* (2014), CERN-ACC-2014-0321
3. F. Gerigk, M. Vretenaar et al., *Linac 4 Technical Design Report*, in *Report* (2006), CERN-AB-2006-084
4. E. Benedetto et al., *Status of the 160 MeV H^- Injection into the CERN PSB*, in *Proceedings, 3rd International Conference on Particle accelerator (IPAC 2012): New Orleans, USA, May 20-25* (2012), p. TUPPR091
5. M. Delonca, C. Maglioni, A. Patapenka, A.S. Martinez, *Internal H^0/H^- Dump for the Proton Synchrotron Booster Injection at CERN*, in *Proceedings, 3rd International Conference on Particle accelerator (IPAC 2012): New Orleans, USA, May 20-25* (2012), p. TUPPR054
6. J.L. Abelleira, W. Bartmann, E. Benedetto, C. Bracco, G.P. Di Giovanni, V. Forte, M. Kowalska, M. Meddahi, B. Mikulec, G. Rumolo, *Painting Schemes for CERN PS Booster H^- Injection*, in *Proceedings, 6th International Particle Accelerator Conference (IPAC 2015): Richmond, Virginia, USA, May 3-8* (2015), p. THPF083
7. A. Lombardi, *Linac4: From Initial Design to Final Commissioning*, in *Proceedings, 8th International Particle Accelerator Conference (IPAC 2017): Copenhagen, Denmark, May 14-19* (2017), p. TUYA1
8. W. Weterings, C. Bracco, R. Noulibos, Y. Sillanoli, P. van Trappen, J Radioanal Nucl Chem **305**, 831 (2015)
9. W. Weterings, C. Bracco, L. Jorat, R. Noulibos, P. van Trappen, AIP Conference Proceedings **1962**, 030003 (2018)
10. NEC, 7540 Graber Road, P.O. Box 620310, Middleton, Wisconsin 53562
11. P. Van Trappen, R. Noulibos, W. Weterings, *Stripping Foil Displacement Unit Control for H^- Injection in PSB at CERN*, in *Proceedings, 15th International Conference on Accelerator and Large Experimental Physics Control Systems (ICALEPCS 2015): Melbourne, Australia, October 17-23* (2015), p. MOPGF121
12. B. Goddard, M. Aiba, C. Bracco, C. Carli, M. Meddahi, W. Weterings, *Stripping Foil Issues for H^- Injection into the CERN PSB at 160 MeV*, in *Proceedings, 1st International Conference on Particle accelerator (IPAC 2010): Kyoto, Japan, May 23-28* (2010), p. THPEB030
13. ACF-Metals, 2239 E. Kleindale Road, Tucson, Arizona, U.S.A
14. GSI, Helmholtzzentrum für Schwerionenforschung GmbH, Planckstraße 1, 64291, Darmstadt, Germany
15. MICROMATTER, Unit #1, 8333 - 130th Street, Surrey, BC V3W 7X4, Canada
16. A. Tatami, KANEKA Corporation, 5-1-1, Torikai-Nishi, Settsu Osaka 566-0072, Japan
17. A. Tatami, M. Tachibana, T. Yagi, M. Murakami, AIP Conference Proceedings **1962**, 030005 (2018)
18. F. Roncarolo et al., *Beam Instrumentation for the CERN LINAC4 and PSB Half Sector Test*, in *Proceedings, 8th International Particle Accelerator Conference (IPAC 2017): Copenhagen, Denmark, May 14-19* (2017), p. MOPAB120
19. S. Burger, *Beam Observation System (BTV) at Stripping Foil for PSB H^- Injection*, in *Engineering Specification 1706232* (2016), `https://edms.cern.ch/document/1706232/1`

Development of a high-density highly oriented graphite stripper

Hiroo Hasebe[1,*] *Hiroki* Okuno[1], *Atsushi* Tatami[2], *Masamitsu* Tachibana[2], *Mutsuaki* Murakami[2], *Hiroshi* Imao[1], *Nobuhisa* Fukunishi[1], *Masayuki* Kase[1], and *Osamu* Kamigaito [1]

[1] Nishina Center for Accelerator-Based Science, RIKEN, 2-1 Hirosawa, Wako, Saitama 351-0198, Japan
[2] Material Solutions New Research Engine, Kaneka Corporation, 5-1-1 Torikainishi, Settsu, Osaka 566-0072, Japan

Abstract. High-density highly oriented graphite sheets provided by Kaneka Corporation have been applied as stripper disks for heavy-ion acceleration at RIKEN RI Beam Factory since 2014. The graphite sheets withstand increasing amounts of beam intensity. We observed the graphite sheets after beam irradiation with an electron microscope and their lifetime was discussed.

1 Introduction

Among many heavy-ion accelerator facilities, static-type carbon foils (C-foils) have been used as charge strippers in heavy ion acceleration for a long time. In recent years, many accelerator facilities have upgraded or plan to upgrade with the aim of providing a higher beam intensity and higher beam energy. Inevitably, the lifetime of the charge stripper foil has become a serious concern for the accomplishment of beam operations. This has led to the replacement of foil strippers, such as carbon and thin metal foils, with gases with low atomic numbers, like He and H_2, or flowing liquid metals.

For uranium (U) acceleration at the RIKEN RI Beam Factory (RIBF), the charge state of U^{86+} is necessary to accelerate U ions up to the final energy of 345 MeV/nucleon. It is difficult to obtain such a high charge state at the stripping energy around 50 MeV/nucleon without C-foil strippers. Before 2011, we used static-type polycrystalline graphite foils with a thickness of 17 mg cm^{-2} fabricated by Arizona Carbon Foil Company [1]. Beryllium disks [2] (with a thickness of 17 mg cm^{-2}) were used from 2012 to 2014. In 2014, we found that high-density highly oriented graphite sheets (GSs), provided by Kaneka Corporation [2, 3], can be applied as the stripper disks for heavy ion acceleration. GS-disks have been used since 2015.

2 Graphite sheet disk stripper

2.1 Characteristics of the GSs

The GSs are prepared from heat-treated polyimide films at temperatures up to 3000°C [3]. The characteristics of the GSs are listed in Table 1. A prominent feature of the GSs is their very high thermal conductivity of 1500 W m^{-1} K^{-1} in the plane direction. In addition, the GSs have a higher thermal diffusivity than copper or aluminum. The GSs have a high density and uniform thickness.

Table 1. Characteristics of the KANEKA GS.

		Units	Typical values
Thermal conductivity	In plane (XY axis)	W m^{-1} K^{-1}	**1500**
	Normal to plane (Z axis)		5
Thermal diffusivity		cm^2 s^{-1}	9.0
Density		g cm^{-3}	**2.0**
Tensile strength		MPa	40
Bending		Cycles	>10000
Electrical conductivity		S cm^{-1}	13000

2.2 GS-disk used in U beam time

The GSs are mechanically strong and can be shaped easily by scissors or a cutter knife. Since the thickness variation of the GS sample was limited, we prepared a new GS-disk by overlapping two sheets so the total thickness became 15 mg cm^{-2}, which is necessary for U beam acceleration. Each sheet, with a thickness of 7.5 mg cm^{-2}, was cut to a circular shape with an outer diameter of 110 mm.

Figure 1 shows a photograph of the four GS-disks after U beam irradiation during the operation period from 2015 to 2017. We do not observe any serious damage by beam irradiation on any disk. Only slight deformation and beam irradiation marks are observed. Disks 1 and 2 were used for two beam times. The beam intensity after the stripper fluctuated slightly at the end of the beam time. Disks 3 and 4 provided an excellent beam during the beam time with no damage and no beam intensity fluctuation, however their use was discontinued after one beam time.

* Corresponding author: hasebe@riken.jp

https://doi.org/10.1051/epjconf/202022901004

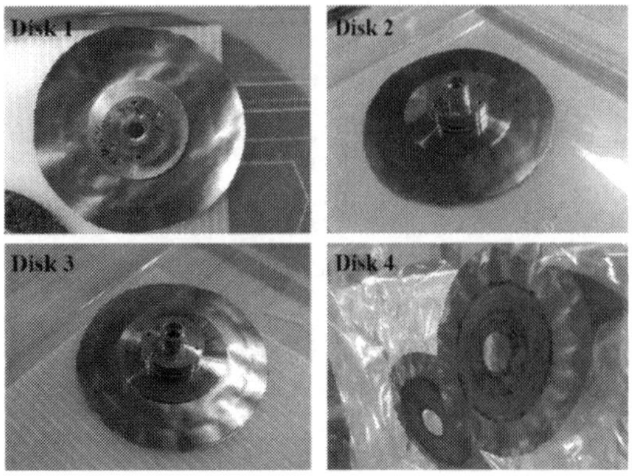

Fig. 1. Four GS-disks after the beam times.

Table 2 summarizes the results of the four GS-disks. The highest beam intensity of 20 eμA was irradiated on Disk 4. However, the intensity downstream did not fluctuate. The total beam particles irradiated on Disks 1 and 2 are shown in Table 2. For Disks 3 and 4, estimations of the irradiated beam amounts are shown. The disks were only slightly deformed. Their actual lifetimes are longer than the shown values. They may be able to cope with a further increase in beam intensity, however, introducing new measures to suppress deformation is necessary in future.

3 Observation scanning electron microscopy

Disk 1 was observed with a scanning electron microscope (SEM) with the electronic scanning micro analyzer (EPMA). A part of the beam irradiation mark was cut with scissors and was placed at the sample stage of the SEM. The surface, the edge surface, and the cross-section were observed in both cases in which the beam was irradiated or not irradiated. To observe the cross-section, a GS was attached to a brass block and was placed upright. As shown in Figure 1, we cannot see any damage by eye to the GS.

3.1. SEM images of the GS surface

Figures 2 (a1) and (a2) show the SEM images of the no-irradiated part, and (b1) and (b2) show the beam-irradiated part. The magnification factors are 1000 times for (a1) and (b1), and 5000 times for (a2) and (b2). In the SEM image with 5000-times magnification, damage of micrometer-order can be seen at the beam irradiated part.

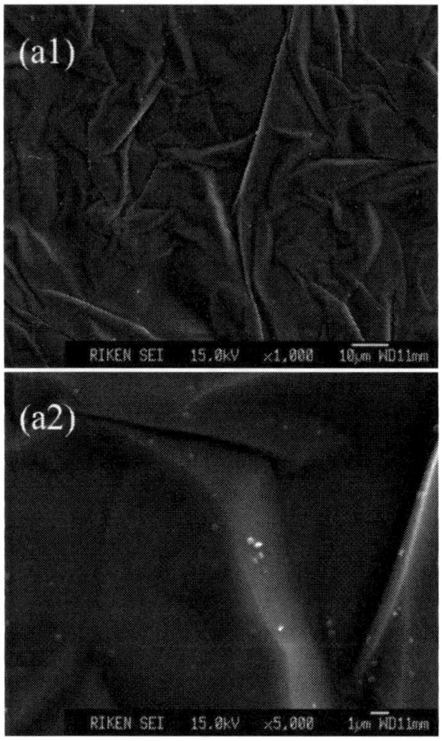

Fig. 2a. SEM images of the GS surface. (a1) and (a2): no-irradiated part.

Table 2. Summary of the results of 4 GS-disks.

Disk name	Maximum Beam intensity (eμA)	Total irradiated particles	Irradiated period (Days)	During irradiation
Disk 1	**17.5**	**2.19×10^{18}**	**40**	**Slight beam fluctuation**
Disk 2	**11.5**	**1.6×10^{18}**	**52**	**Slight beam fluctuation**
Disk 3	**15.8**	**1.2×10^{18} (Estimation)**	**25**	**No beam fluctuation**
Disk 4	**20**	**2.5×10^{18} (Estimation)**	**43**	**No beam fluctuation Deformed a little**

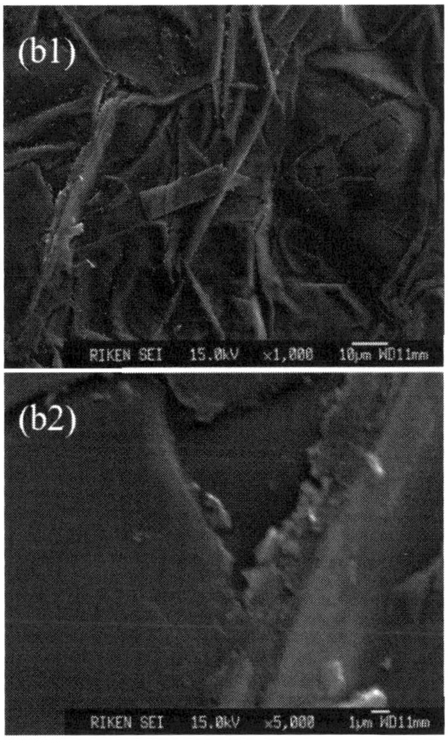

Fig. 2b. SEM images of the GS surface. (b1) and (b2): beam-irradiated part.

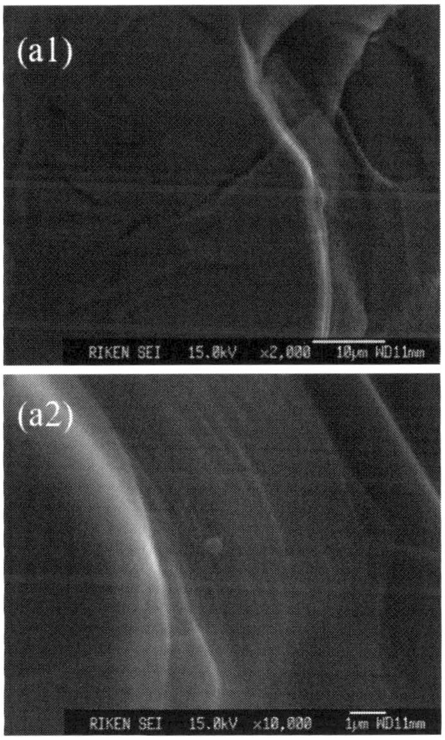

Fig. 3a. SEM images of the GS surface edge taken from above. (a1) and (a2): no-irradiated part.

3.2 SEM images of the GS surface at the edge from above

Figures 3 show the SEM images taken of the GS surface edge from above. Figure (a1) and (a2) show the no-irradiated part, and (b1) and (b2) show the beam-irradiated part. The magnification factors are 2000 times for (a1) and (b1), and 10,000 times for (a2) and (b2). As well as the surface, you can also see damage of micrometer-order at the edge of the beam-irradiated part. A smooth multilayer structure is observed at the no-irradiated part.

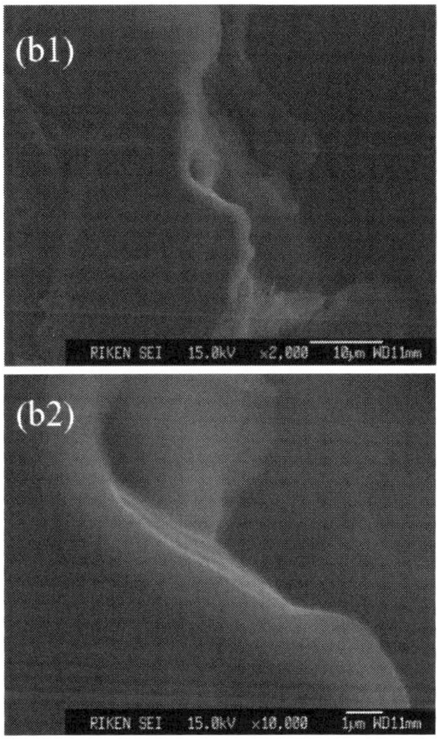

Fig. 3b. SEM images of the GS surface edge taken from above. (b1) and (b2): beam-irradiated part.

3.3 SEM images of the GS cross-section

Figures 4 show the SEM images observing the GS cross-section. Figure (a1) and (a2) show the no-irradiated part, and (b1) and (b2) show the beam-irradiated part. The magnification factors are 2500 times for (a1) and (b1), and 10,000 times for (a2) and (b2). You can also see damage of micrometer-order at the center of the irradiated part. However, we cannot observe the change in the thickness. To investigate the carbon purity, EPMA analysis was also performed in both cases for the beam irradiated and no-irradiated parts. Substances other than carbon were not detected at any part.

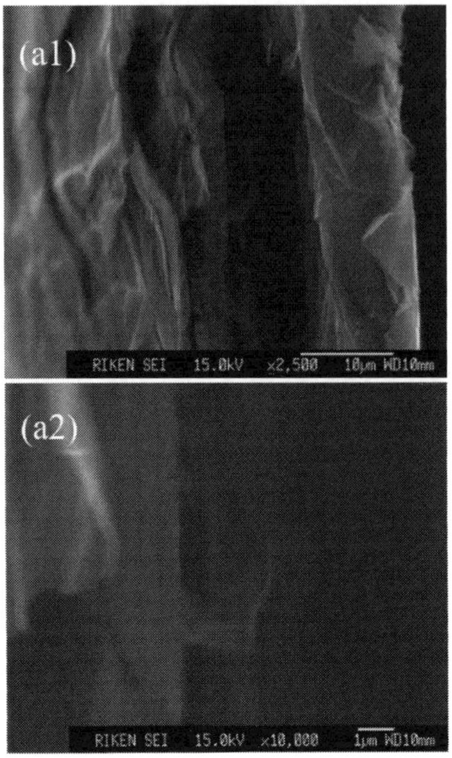

Fig. 4a. SEM images of the GS cross-section. (a1) and (a2): no-irradiated part.

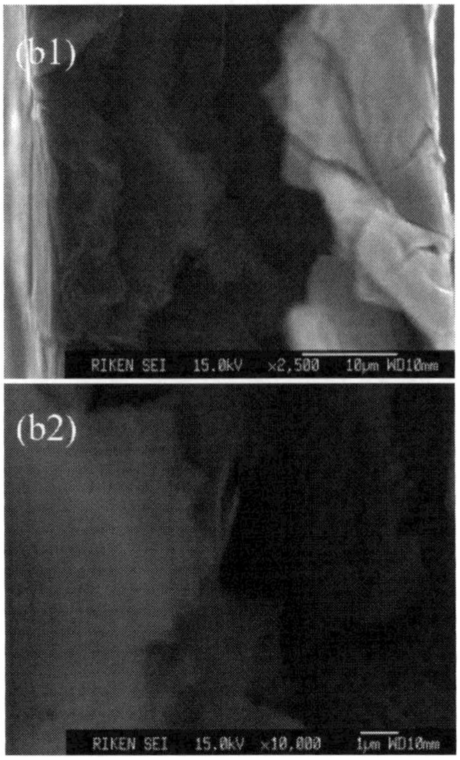

Fig. 4b. SEM images of the GS cross-section (b1) and (b2): beam-irradiated part

4 Summary

We have been using KANEKA GSs as the second stripper for uranium acceleration since 2015. This GS stripper works successfully and provides stable high-intensity beams for a long period. Currently, the used GS disk is replaced after one beam time operation due to the increase in beam intensity. Used GSs seem to have no beam-irradiation damage when observed visually. However, when observed using a SEM, damage of micrometer-order is observed on the surface and inside the GS. The reason for the damage may be due to weakening of the bond strength or lattice defects caused by the beam irradiation. It is noticed that the thickness of the GS is not changed by beam irradiation at the present intensity. Further analysis by Raman spectroscopy and TEM may resolve details. It is concluded that KANEKA's multi-layer GS is a suitable material for use as charge strippers.

References

1. ACF-Metals Arizona Carbon Foil Co. Inc., URL: http://www.techexpo.com/firms/acf-metl.html
2. H. Hasebe, H. Okuno, H. Kuboki, H. Imao, N. Fukunishi, M. Kase, O. Kamigaito, AIP Conference Proceedings **1962**, 030004 (2018)
3. A. Tatami, M. Tachibana, T. Yagi, M. Murakami, AIP Conference Proceedings **1962**, 030005 (2018)

The irradiation study of carbon stripper foils by a 12 MeV Ar beam

Wen Jiang[1,*], *Wolfgang* Mittig[2], *Frederique* Pellemoine[3], *Tan* Ahn[4], *Daniel* Robertson[4], *Edward* Stech[4], and *Nalin* Kumar[5]

[1]Purdue University, Nuclear Engineering Department, IN 47907-2017, USA
[2]Michigan State University, National Superconducting Cyclotron Lab, MI 48824, USA
[3]Michigan State University, Facility for Rare Isotope Beams, MI 48824, USA
[4]University of Notre Dame, Department of Physics, IN 46556, USA
[5]UHV Technologies, Inc., TX 76104, USA

Abstract. Irradiation performance of multilayer HBC foils which layers are more than 20 were conducted for the first time, and different types of carbon stripper foils were dosed under a series of irradiation conditions systematically. Eight (8) types of carbon stripper foils were striked by Ar^{4+} beam under different intensities and fluences. Multilayer foils performed generally better than the monolayer ones. The existence of boron layer helped to increase the resistance of the foils to irradiation damage, elongating their lifetimes.

1 Introduction

Carbon stripper foils belong to one of the dominant types of stripper foils and have been studied for decades [1]-[5]. Diamond like carbon (DLC) and hybrid boron doped carbon (HBC) foils are two popular kinds among all carbon stripper foils. Multilayer HBC foils have been considered for their long lifetime, and attracted much interest [6-10]. The preparation method of the HBC foils has been the subject of many studies, yet the relationship between the characteristics of the layers and the lifetime of the foils has not yet been systematically studied. The irradiation performance of foils with more than 20 layers has not yet been reported. For such foils with thickness around 2μm or less, where each layer is less than 100nm thick, some nanoscale effects [11-12] are expected to take place. In this paper, 8 types of carbon stripper foils, including foils of 48 layers and 100 layers, were irradiated with a Ar^{4+} beam of 12 MeV under different intensities and fluences for the first time. The irradiation was performed with the 5U accelerator of University of Notre Dame.

2 Materials

The stripper foils used in this study were all supplied by UHV Technologies, Inc.. There were totally eight types of foils: 2 types of monolayer DLC foils, 10 wt% boron mix 90 wt% carbon monolayer foil, and 7 layers, 9 layers, 17 layers, 48 layers, 100 layers HBC foils. The layout of the foils is listed in Table 1, and each type of foil was assigned a number, for the convenience of discussion. The index of the foils is showing in Table 1.

Table 1. Types of the foils. The nominal thicknesses were between 2-2.5μm (around 500 μg/cm²), except the 10% boron mix 90% carbon monolayer foil, which thickness is 1.6μm.

Num.	Foil type/layout	Composition
1	One layer DLC, type 3	Amorphous carbon
2	One layer DLC, type 1	Amorphous carbon
3	7 layers HBC, 4B+3C	6% B + 94% C
4	9 layers HBC, 5B+4C	20% B + 80% C
5	17 layers HBC, 9B+8C	20% B + 80% C
6	48 layers HBC, 24B+24C	83% B +17% C
7	100 layers HBC, 50B+50C	83% B + 17% C
8	One layer mixture	10% B+90% C

3 Experiment

All the foils were mounted on brass frames, glued by either AquaDAG ® (graphite coating) or double-sided graphite tape. The effective irradiation area of the foils is 6.35×5.84 mm². The frames were further mounted on copper plates to be irradiated. Each plate bears 4 different foils, that can be irradiated under the same condition. Thus the 8 types of foils were mounted on 2 plates for each run's condition. The plates were then loaded on a target holder with water-cooling system. See Fig. 1. The spot size was adjusted to be large enough to cover all 4 foils and it was re-defined each time the beam intensity was changed, or after several rounds had been done.

* Corresponding author: jiang568@purdue.edu

(a) (b)

Fig. 1. Sample assembly. (a) Samples loaded on holder. Water goes through the bottom flange, behind the copper plate; (b) Irradiation line. The 2 plastic tubes belong to water cooling system.

Before irradiation, plastic sheets were irradiated to confirm the beam size. The beam size was re-measured every time the beam intensity was changed, or after several runs had been done. Alpha particle attenuation was used to determine the thickness of the foils. The alpha source used was Am-241, 3μCi.

4 Irradiation of the foils

Ar^{4+} beam had an energy of 12 MeV. Multiple beam intensities and fluences were used. The initial intensity we aimed at was 10μAe for Ar^{4+}. Yet to prevent extreme damage of the foils before reaching higher fluence, we had to use lower intensities. One example of the extremely damaged samples after irradiation is shown in Fig. 2.

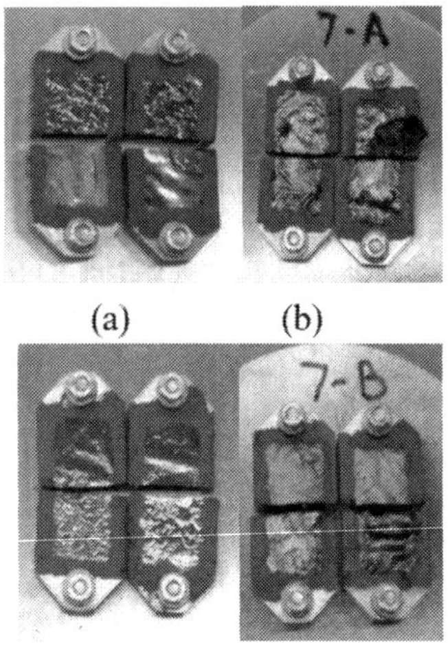

(a) (b)

(c) (d)

Fig. 2. Foils before and after irradiation. (a) (from left to right, up to bottom) 9 layers HBC, 7 layers HBC, DLC type1, DLC type3 foils before irradiation; (b) 9 layers HBC, 7 layers HBC, DLC type1, DLC type3 foils after irradiation; (c) (from left to right, up to bottom) 17 layers HBC, 48 layers HBC, 100 layers HBC, B+C mixture foils before irradiation; (d) 17 layers HBC, 48 layers HBC, 100 layers HBC, B+C mixture foils after irradiation.

Pictures were taken each time after irradiation, the damage of all the foils tested is evaluated based on the pictures. The damages were classified to 6 grades:

- 0-No damage;
- 1-No visible cracks or peelings, but uneven surface appeared;
- 2-Ralatively even surface, with visible cracks or peelings;
- 3-Local deformation, with small scale cracks or peelings;
- 4-Badly damaged, with severe cracks or peelings (may accompany with large deformation, depends on foils);
- 6-Foil failed and removed.

Three variables were used for evalutaion: intensity, fluence and the type of foil. Upon a correlation analysis, we showed that the beam intensity is a less significant factor with respect to the foils' damage. Therefore, the average damage under certain magnitudes of fluence were taken for each type of foil regardless of beam intensity. The result is shown in Fig. 3.

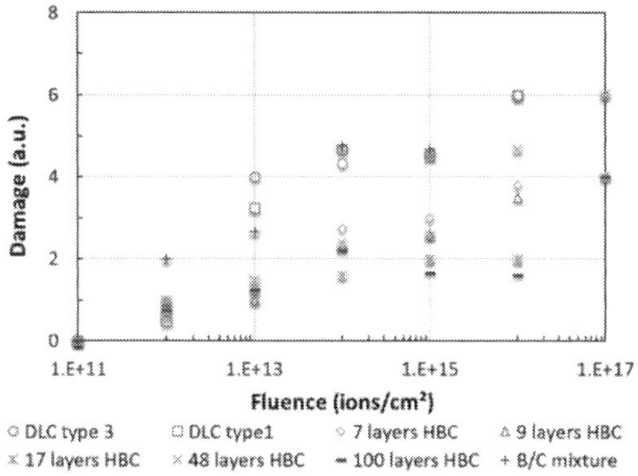

Fig. 3. The damage degree of 8 types of foils under different magnitudes of ion fluence.

It can be seen that below 10^{13} ions/cm^2, all the foils are "safe", but above this fluence, DLC foils and B/C mixture foil become heavily damaged (average grade >4) and failures start to appear above 10^{14} ions/cm^2.

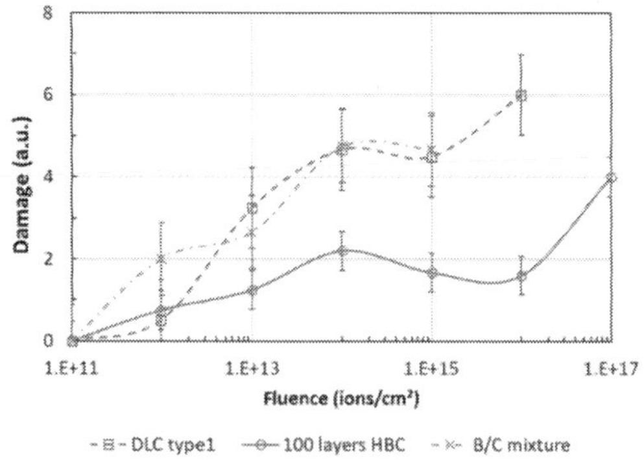

(a)

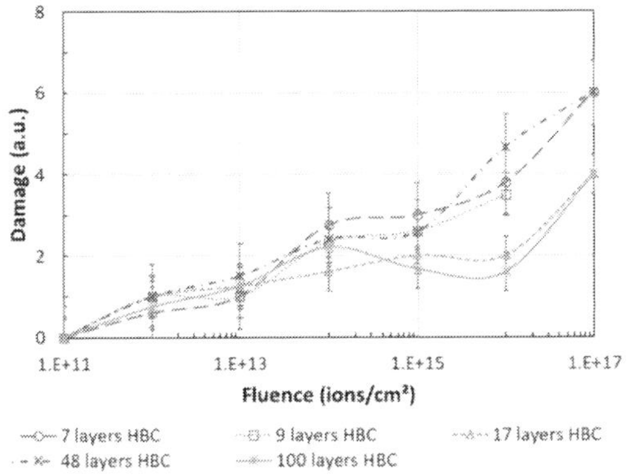

(b)

Fig. 4. Comparison of the damage grade with respect to fluence of two specific sets of samples (a) DLC, HBC and B/C mixture (b) HBC foils of different number of layers.

Specifically, two sets of samples are compared: one DLC foil, one HBC foils, and the B/C mixture; and foils of various number of layers (See Fig. 4 (a) and (b) respectively). The error bars correspond to the deviation of foils struck by beams with different intensities. From Fig. 4. (a), it can be observed that DLC foil and B/C mixture foil became damaged rapidly above 10^{13} ions/cm², and are both near to failure at 10^{14} ions/cm². On the contrary, 100 layers HBC foil remained relatively undamaged up to 10^{17} ions/cm². From Fig. 4.(b), one can find that below 10^{14} ions/cm², the performance of all the foils is similar, and starts to deviate after 10^{14} ions/cm². Despite no obvious relationship between damage degree and the number of layers, it is not until 10^{16} ions/cm² when some foils start to fail. Surprisingly, 17 layers HBC behaved almost the same as 100 layers one

5 Swelling study of the irradiated foils

To quantify the irradiation performance of the foils, one method was to characterize the swelling of the foils. This parameter can be deduced from the thickness of the samples using equation (1).

*Swelling=(thickness of the irradiated samples-thickness of the unirradiated sample)/thickness of the unirradiated sample*100% (1)*

The thickness of the foils was determined by the Energy loss of Alpha-particles method using the 5480 keV alpha energy of an Am-241 source detected by a Si detector with a full width at half maximum (FWHM) of 47 keV.

Stopping and range of ions in matter (SRIM) was used to evaluate the thickness of all sorts of foils. For the α particles of Am-241 in carbon foils, dE/dx= 783.27 keV/mg/cm² considering a density of the foils as 2.25 g/cm³.

100 layers HBC and DLC (type 1) foils irradiated under different conditions were measured. The analysis of the measurements is found in table 2 and table 3 respectively where thickness (in μm) and swelling (in %) are given. The result indicates that the thickness of foils irradiated under different conditions are not much changed.

Table 2. Analysis of the measurements of 100 layers HBC foils

Irradiation fluence (ions/cm2)	Irradiation intensity (μAe)	Thickness evaluated by SRIM (μg/cm2)	Thickness (μm)	Swelling (%)
0	0	261.03	1.16	
1.00*10^15	0.23	267.34	1.19	2.29
1.40*10^12	1.4	263.83	1.17	0.95
1.40*10^13	1.4	252.12	1.12	-3.53
1.40*10^14	1.4	267.87	1.19	2.49
7.01*10^14	1.4	259.19	1.15	-0.83
1.40*10^15	1.4	269.01	1.19	2.93
1.00*1016	4	262.87	1.17	0.58
3.00*10^12	10	267.52	1.19	2.36
1.55*10^15	10	259.25	1.15	-0.80

Table 3. Analysis of the measurements of DLC foils.

Irradiation fluence (ions/cm2)	Irradiation intensity (μAe)	Thickness evaluated by SRIM (μg/cm2)	Thickness (μm)	Swelling (%)
0	0	585.08	2.60	
1.40*10^12	1.4	600.33	2.67	2.38
1.40*10^13	1.4	593.16	2.64	2.32
1.40*10^14	1.4	584.61	2.60	-0.30
7.01*10^14	1.4	588.14	2.42	0.01
1.40*10^15	1.4	569.70	2.53	-2.89
5.02*10^15	1.4	580.15	2.58	-0.80
1.40*10^16	1.4	544.97	2.42	-5.78
1.55*10^15	10	532.39	2.36	-8.15

5 Discussion

As discussed in section 4, the multilayer HBC foils generally had a longer lifetime than DLC foils. The B/C mixture foils performed worst among HBC foils, even worse than DLC foils sometimes. The 100 layers HBC foil had the best resistance toward irradiation, yet the lifetime of other multilayer HBC foils did not show a regular trend with increasing number of layers as expected. In many cases, the 7 and 17 layers foils had longer lifetime than 48 layers ones. This is particularly eminent for 17 layers foils, which behave almost identical to the 100 layers ones.

The 100 layers foils performed best among all the tested foils, also with respect to their structural steadiness. When radiation damage appears, most of the other multilayer foils exhibit layer debonding, while the layers of 100 layers foils, despite the top layer peels sometimes, stayed essentially undamaged even under high intensities or high fluences. A comparison between the 7 and 100 layers is shown in Fig. 5. The pictures were taken by a microscope.

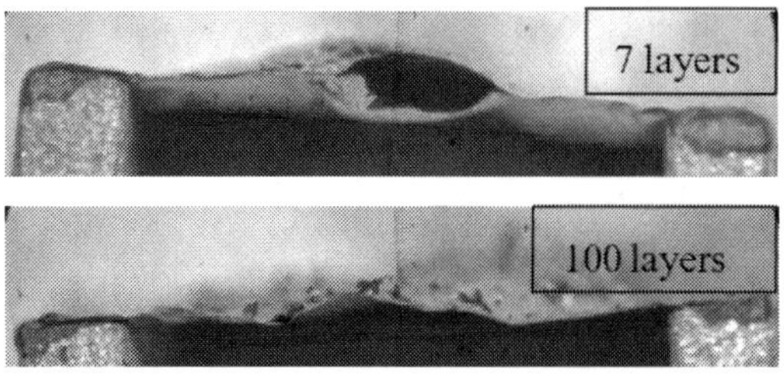

Fig. 5. A view of the border of 7 layers and 100 layers HBC foils after an irradiation of a fluence of $7*10^{14}$ ions/cm^2 at 1.4 μAe.

These observations made it reasonable to believe that there is a connection between the firm layer structure of 100 layers foils and their long lifetime. All the multilayer foils were produced by the same technique, Pulse-Laser Deposition. Therefore, it is less possible to be the processing which made the difference. We tried to explore the reason for the outstanding performance of 100 layers foils from the microstructure's side. Table 2 indicates that the total thickness of this type of foil is around 1 μm, such that the thickness of each layer is about 10 nm. The carbon layers of all the multilayer HBC foils used were supposed to be consisted of DLC structure. It was suggested that upon irradiation, the sp3 bond of DLC tend to transform to sp2 bonds, possibly forming nanographite structures [11, 12]. The displacement of carbon atoms due to energetic particles, heavy ions here particularly, can subsequently occur. As a result, interstitial defects will cause crystallite growth perpendicular to the layer planes, whereas coalescence of vacancies will cause a shrinkage parallel to the layer planes [13]. This crystallite dimensional changes leads to the generation of new porosity [13], which has the potential to become the source of cracks. This can explain the initiation of the cracks in the foils in Fig. 2. The interstitial defects, one form of Frenkel defects, and crystallite distortion discussed above are also known as "Wigner Effect", which is an important issue of graphite when subjected to high energy irradiation[14], resulting in energy storing through lattice swelling. This energy storage to nanodiamond is even higher [15]. At high temperature (>600°C), the defect can be annealed, while below 600°C the distortion of lattice grows as the temperature increases [14]. Without temperature controlling system, the temperature within our samples during irradiation rose as the irradiation intensity increased. The highest temperature (at 10 μAe) was estimated to be 670°C, when the highest degree of lattice distortion occurs-- further explains the derivation of cracks. In terms of multilayer structure, it is reasonable to believe that boron atoms can act as interstitials which diffuse into diamond lattice and recombine with vacancies, forming a stable interface tolerant to irradiation. This may be similar to the promotion of radiation tolerance in metal/metal and metal/oxide interfaces due to the point defects (interstitials, vacancies) recombination [15]. This hypothesis may explain the prominent performance of 100 layers HBC foils. With the increase of thickness of carbon layer in fewer layers foils, the Wigner Effect is expected to be dominant, and the resistance to irradiation declines. Even so, the multilayer foils still have longer lifetime than monolayer ones, including B/C mixture foil, proving that boron layers increase the resistance of the DLC layers toward irradiation.

According to section 5, it is easy to find that the thickness of 100 layers HBC foils irradiated under different conditions did not change too much. Considering the uncertainty, the thickness of DLC foils didn't change too much either, despite a decreasing tendency seemed to appear for DLC foils at higher fluences, see figures 6 and 7.

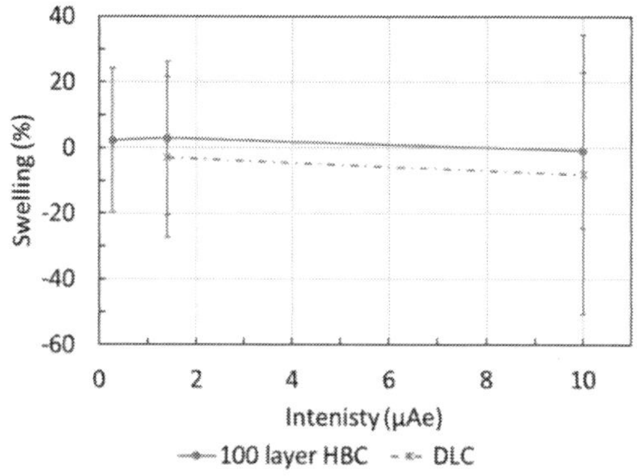

Fig. 6. Swelling of 100 layers HBC and DLC foils irradiated under 0.23 (N/A for DLC foils), 1.4, 10 μAe intensities and the same fluence of $1.40*10^{15}$ions/cm2.

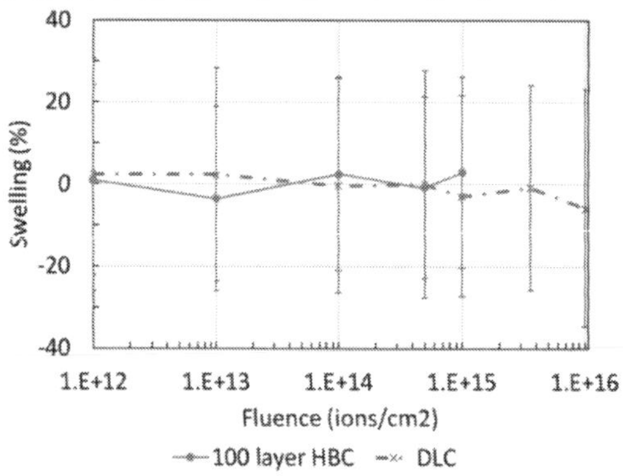

Fig. 7. Swelling of 100 layers HBC foils irradiated under 1.4 μAe intensity and the increasing fluence.

It is worth noticing that the swelling here indicates the increasing in thickness, yet in real case, the thickness of the foils should reduce when the foils "swell", because the expansion of the surface area is accompanied by the attenuation in thickness, when the bulk density remains the same.

The expansion of the surface area of the foils is majorly caused by thermal effect [16, 17]. On the other hand, the thickness could reduce due to the "ion hammering effect" [18], which has an outcome of swelling. Yet the reduction in measured thickness does not always indicate swelling. It is very possibly due to the sputtering of surface atoms or the cracks on the foil as well. Therefore, thermal expansion and "ion hammering effect" are not enough to explain the thickness change of the foils. The thickness measurements give an effective thickness, which is a combination of the "true thickness" with the angle of the ripples (deformation), especially for irradiated DLC foils. This argument is also indicated by the high uncertainty of the "swelling" percentage.

Despite the complication discussed above, it is still possible to tell by comparing the alpha particle test results of 100 layers HBC foil and DLC foil that multilayer foils maintain better their sizes than DLC foils and are more resistant to irradiation. This result is coherent with the conclusion made earlier.

6 Conclusion

In the present study, irradiation performance of multilayer HBC foils with more than 20 layers were observed for the first time, and different types of carbon stripper foils were systematically irradiated under a series of irradiation conditions. 8 different types of carbon stripper foils were irradiated using the Ar^{4+} 12 MeV beam from the 5U accelerator of the University of Notre Dame. It was found that below 10^{13} ions/cm^2, regardless of beam intensity, all the foils are "safe". Then non-multilayer foils start to show heavy damage and eventually fail. Among all the foils, 100 layer HBC foils performed best upon irradiation, not only having longest lifetime (10^{17} ions/cm^2 under 10 μAe), but also having the most steady multilayer structure. This is believed to be the reason for the better radiation resistance. Multilayer foils performed generally better than the monolayer ones. Boron layer may help to increase the resistance of the foils to irradiation damage, elongating their lifetimes. The swelling of the irradiated DLC type 1 foil and 100 layers HBC foil were measured via alpha particle attenuation. Although the "swelling" here is caused by complicated factors, it can still be concluded that the 100 layers HBC foils is relatively stable in size upon irradiation, and the swelling of DLC foils decrease with the increasing of fluence at 1.4 μAe intensity.

The author would like to give thank to John Yurkon from NSCL, MSU, for his help in training to use and setting up the equipment used in this work.

References

1. N.R.S. Tait, Nucl. Instrum. Methods **184**, 1, 203-210 (1981)
2. G. Dollinger, P. Maier-Komor, Nucl. Inst. Methods Phys. Res. B **53**, 352-354 (1991)
3. I. Sugai, M. Oyaizu, H. Kawakami, C. Ohmori, T. Hattori , K. Kawasaki, M. J. Borden, R. J. Macek, Nucl. Instrum. Methods Phys. Res., Sect. A **362**, 70-76 (1995)
4. H. Muto, M. Oyaizu, K. Kawasaki, T. Hattori, Radiation Effects and Defects in Solids **164**, 9, 561–565 (2009)
5. W. Barth, M.S. Kaiser, B. Lommel, M. Maier, S. Mickat, B. Schlitt, J. Steiner, M. Tomut, H. Vormann, J. Radioanal Nucl. Chem, **299**, 1047-1053 (2014)
6. I. Sugai, Y. Takeda, M. Oyaizu, H. Kawakami, T. Hattori, K. Kawasaki, Japanese J. Appl. Phys., **45**, issue 11, 8848–8852 (2006)
7. S.K. Zeisler, V. Jaggi, Nucl. Instrum. Methods Phys. Res., Sect. A, **590**, 18-21 (2008)
8. I. Sugai Y. Takeda, M. Oyaizu, H. Kawakami, Y. Irie, Y. Arakida, K. Hara, H. Hattori, K. Kawasaki, J. Kamiya, M. Kinsho, K. Kuramochi, Nucl. Instrum. Methods Phys. Res., Sect. A, **561**, 16-23 (2006)
9. S.K. Zeisler, V. Jaggi, Nucl. Instrum. Methods Phys. Res., Sect. A, **613**, 434-435 (2010)
10. V. Jaggi, S. Olson, B. Kumar, S.K. Zeisler., J Radioanal Nucl Chem, **299**, 1019-1021 (2014)
11. H.C. Chen, U. Palnitkar, B. Sundaravel, I-Nan Lin, A. Pratap Singh, R. Kumar Ranade, Diam. Relat. Mater., **18**, issues 2-3, 164–168 (2009)
12. C. Popescu, G. Dorcioman, B. Bita, C. Besleaga, I. Zgura, C. Himcinschi, A. C. Popescu., Appl. Surf. Sci. **390**, 236-243 (2016)
13. L.L. Snead, T.D. Burchell, Y. Katoh, J. Nucl. Mater. **381**, 55–61 (2008)
14. S. Tsai, E. Huang, J. Kai, F. Chen. J. Nucl. Mater. **434**, 17-23 (2013)
15. V. Shutthanandan, S. Choudhury, S. Manandhar, T. C. Kaspar, C. Wang, A. Devaraj, B. D. Wirth, S. Thevuthasan, R. G. Hoagland, P. P. Dholabhai, B. P. Uberuaga, R. J. Kurtz, Adv. Mater. Interfaces **4**, 14, 1700037 (2017)
16. I. Sugai , Y. Takeda, M. Oyaizu, H. Kawakami, Y. Irie, A. Takagi, H. Hattori, K. Kawasaki., Nucl. Instrum. Methods Phys. Res., Sect. A, **613**, 457–461 (2010)
17. H. Hasebe, H. Okuno, H. Kuboki, H. Ryuto, N. Fukunishi, O. Kamigaito, A. Goto, M. Kase, Y. Yano, Nucl. Instrum. Methods Phys. Res., Sect. A, **613**, 453–456 (2010)
18. J.A. Nolen, F. Marti, Reviews of Accelerator Science and Technology, **6**, 221-236 (2013)

Methods of targets' characterization

Christelle Stodel[1,*]

[1]Grand Accélérateur National d'Ions Lourds (GANIL), CNRS/CEA, Bvd Henri Becquerel, BP 55027, 14076 Caen CEDEX 5, France

Abstract. The paper reviews the characterization's techniques for solid targets used in nuclear physics with special emphasis on actinide targets. The determination of the thickness, isotopic and chemical composition are described for actinide targets during their fabrication process. Accent is given on their monitoring during irradiation.

1 Introduction

In nuclear physics, targets are an essential ingredient for the success of experiments. Precise knowledge of target properties like e.g., the thickness, its distribution (or homogeneity) over the surface and its purity either isotopically or chemically is irrevocable for proper analysis of experimental data.

Indeed, knowledge of the kind and level of target impurities is required for evaluating side reactions competitive to the one under study and in estimating the background level. In addition, the characterization of actinide targets is important as the base material consists often of various radioactive isotopes, including decay products. Although some nuclides are present in a very low amount, they can induce high activity due to their short half-life but also they might interfere the measurements. For instance, in the thermal-neutron induced fission reaction of ^{238}U target cited in [1], the main contributions (at 37.3% and 5.3 % respectively) in the measured cross-section comes from the high fissility of ^{235}U present at only 12 ppm in the raw material and from ^{239}Pu formed after neutron capture in ^{238}U.

The areal density (thickness) of targets and stripper foils is also a crucial parameter for nuclear physics experiments. As a matter of fact, it is taken into account in the kinematical characteristics of the produced nucleus of interest, which will be used to tune correctly the spectrometer in order to optimize its transmission. Moreover, the thickness of the targets influence directly the results of cross-sections measurements which are often required with uncertainties in the level of percent. P. Schillebeeckx et al. [2] and D. Sapundjiev et al. [3] report that the improvement in the quality of cross-section data in various fields (astrophysics, neutron-induced reaction for nuclear power production, fuel cycle) relies mainly in the high quality targets with known properties at any time. Another aspect to be considered is the variation of target properties during particle irradiation due to physical and chemical instability of the material, then the monitoring of various target performance parameters all along the experiments is essential.

This paper does not pretend to be exhaustive and read as a textbook, but recalls some general principles and results of the used techniques for which details are provided in the quoted bibliography. As mentioned, only active material and monitoring under irradiation are under consideration, additional information on target characterization can be found in [4-6] and references therein.

The section 2 describes specific techniques for thickness, homogeneity and purity characterization applied to active targets during their fabrication process. The section 3 reports on methods performed during the irradiation of the material in order to monitor their thickness.

2 Characterization of actinide targets

2.1 Thickness and homogeneity

For radioactive targets, their activity at a certain time, $A(t)$, determines directly the number of atoms, $N(t)$, of a sample through its definition as $A(t) = \lambda N(t)$ knowing the decay constant, λ, of the radio-nuclide under consideration. The obtained areal density given in atoms/cm² is then straightforward applicable in the cross-sections equation.

2.1.1 Low geometry α-particle counting (LGA)

Counting at a defined solid angle is the most accurate method for activity measurements of alpha emitters, and is also suited for x-ray emitters. The method consists of measuring with a silicon detector, the α-particle recoiling out of the active sample. Namely, the counting particles move undisturbed along a straight line from the source to the detector, other particles, i.e. those out of the opening solid angle or scattered ones are absorbed in the diaphragm placed just before the detector or in side

* Corresponding author: stodel@ganil.fr

materials (e.g. threads or baffles). All geometric parameters (distance source-detector, knife-edge aperture of the diaphragm, sample size) define precisely the solid angle and the silicon background is as low as possible. With LGA method the uncertainties as low as 0.02 % are feasible [7, 8]. As reported in [9], two different set-ups can be used according to the activity of the samples, either with a variable or a fixed source-detector distance (up to 20 cm or 1.6 m under glove box respectively) where uncertainties of 0.3-0.5 % are reached with tens of spectra acquired in about 40 hours. The distribution of the activity over the sample area is achieved by introducing an additional diaphragm other the source and moving it [10].

2.1.2 Radiographic Imaging (RI)

For large area targets (10-30 cm^2 area), Radiographic Imaging is well suited for activity measurements. Indeed, the active target is exposed in front of a phosphor screen called imaging plate (IP) which traps and stores the radiation energy. Then the IP is scanned with a laser beam which releases the energy as luminescence. This "photo-stimulated luminescence" (PSL) phenomenon is thus proportional to the amount of radiation exposed enabling to investigate the homogeneity of the target layer. The resulting spatial resolutions depend on the radioactive decay properties of the material: less than 200 μm for ^{198}Au ($E_{\beta-}$ = 961 keV, E_γ = 412 keV) [11, 12], 42 μm was achieved for ^{238}U (E_α = 4151 keV, E_γ = 50 keV) [13]. In order to determine the absolute target activity a calibration with a reference sample is needed [12] and a thickness standard deviation of ±5.2% is obtained. This process requires about 6 hours per target, it can be applied before and after irradiation as shown in [14]. As mentioned in [15], when test experiments are performed with homologous of the actinide elements, radiotracers are used enabling the application of RI for layer homogeneity measurements.

2.2 Isotopic abundances

2.2.1 Thermal Ionization Mass Spectroscopy (TIMS)

Thermal Ionization Mass Spectrometry (TIMS) is a highly sensitive technique to measure isotope ratio of nuclear material. It combines the thermal ionization effect of the sample to be analysed with a separation corresponding to the mass-to-charge-ratio of each isotope as performed with an electromagnet [16]. This method is applied to an aliquot of the active solution used for the electro-deposition. The atomic abundances of the solution are precisely measured with a dynamic range of the ratios up to 10^6. The activity fractions of each isotope are then deduced and when applied to the total activity measured with LGA technique, the areal density of each active element in the target is deduced [9, 10].

2.2.2 Neutron Activation Analysis (NAA)

Neutron Activation Analysis (NAA) is an analytical technique based on gamma spectroscopy measurements of samples irradiated with neutron in nuclear reactors. This method is based on the identification by measuring the gamma emission of the reaction products ${}^{N+1}_{Z}A$ of a neutron-induced reaction ${}^{N}_{Z}A(n,\gamma){}^{N+1}_{Z}A$, with ${}^{N}_{Z}A$ the isotope present in the supernatant electrolyte solution to be identified. In order to be unbiased by the total neutron flux, a reference sample can be co-irradiated under similar conditions to compare the activities of the unknown sample (A_x) and a known standard (A_{st}), the element mass (m) is then obtained by : $m_x = m_{st}(A_x/A_{st})$ [17]. The significant advantage of NAA is simultaneous determination of up to 30-40 elements with remarkably low detection limits, especially particular lanthanides and actinides can be evaluated in the range of 0.1-10 ppb due to their high cross-section for thermal neutron capture [18]. The process can be considered as time consuming (6 – 24h) compared to other analytical techniques [12] but fits well in cases when a direct radiometric measurement of the target is not possible since (i) the activity is far too high, (ii) in case of α-particle measurements the layer is too thick, or (iii) lateral dimensions of the target do not allow the introduction of the sample into a counter. The NAA technique is applied at Mainz to qualify actinide targets. Fig. 1 displays the γ-ray spectrum of 1 ml of ^{244}Pu supernatant solution (extracted in the deposition cell after deposition) irradiated for 2 hours with a thermal neutron beam from TRIGA Mainz reactor at a flux of 10^{11} cm^{-2} s^{-1}. After a cooling time of 0.5 hour, the γ-ray measurements are performed for 10 minutes, main γ-rays correspond to ^{245}Am and ^{245}Pu.

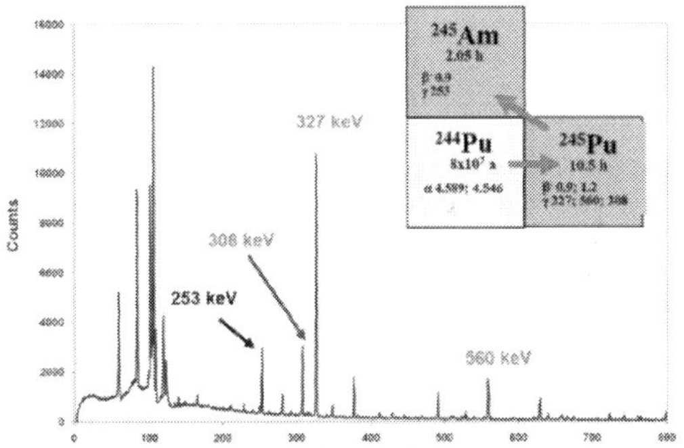

Fig. 1. γ-ray spectrum of a ^{244}Pu solution irradiated at the research reactor TRIGA Mainz [19].

2.3 Chemical identification

2.3.1 Rutherford Back Scattering (RBS)

The thickness of the deposited material and the backing as well as the composition and concentration of the materials can be assigned with the Rutherford Back Scattering (RBS) method [20, 21]. A beam of particles or light ions with energy below the Coulomb barrier is impinging on targets, then scattered projectiles are registered by detectors placed at various angles to the incident beam. In order to estimate the homogeneity of the target, the measurement can be performed at different position on the surface. As shown in figure 3 of [22], the spectrum reflects the homogeneity of the deposit (long tail of the uranium low energy peak) and the presence of oxygen.

2.3.2 Energy Dispersive X-Ray Spectroscopy (EDS)

The chemical composition of deposited layer is obtained with energy dispersive X-ray spectrometry (EDS) using an electron beam and X-ray detectors measuring the LM-transition line of present elements [12]. This method is often coupled with a Scanning Electron Microscope in order to investigate the morphology of the targets [13].

2.3.3 X-Ray Photoelectron spectroscopy (XPS)

More detailed analysis of the composition and binding of the deposition is performed with X-ray photoelectron spectroscopy (XPS). It consists of scanning the target's surface with micro-focused mono-energetics X-ray source and of analysing the emitted electrons. From the shape, position and intensity of photoelectric peaks, one can deduce the chemical state and quantify the element [15, 13]. Impurities in the range of percent are analysed with the EDS and XPS methods.

3 Monitoring of targets under irradiation

3.1 Electron attenuation

By comparing the emitted current of an electron beam of 20 or 30 kV with the scattered one passing through a material, a relative thickness measurement of targets is obtained [23, 24]. When targets are mounting on a wheel, the relative thickness information over the target area is achieved by deflecting the electron beam in the radial direction (with a magnetic deflector) across the rotation, as depicted in Fig. 2.

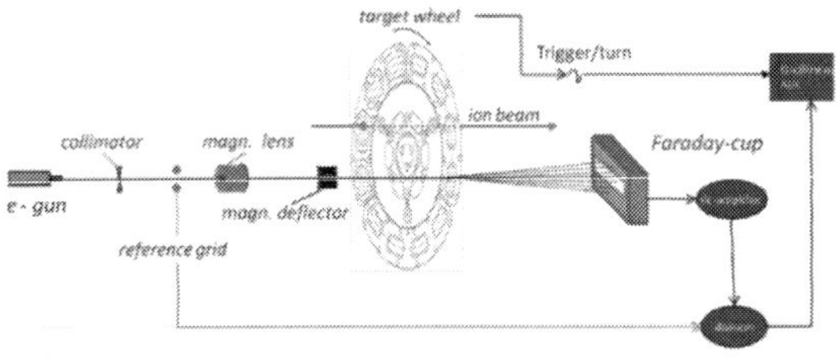

Fig.2. Scheme of the electron gun with associated electronics and target wheel [25].

Due to the narrow width of the electron beam, the position resolution is about 0.5 mm and the accuracy in the determination of the relative thickness is ±2%. As the electron beam, delivered by an electron gun, is parallel to the heavy ion beam used for the nuclear reaction, the targets can be analysed during irradiation, and the thickness mapping lasts at least 100 times the period of the wheel in order to achieve a spatial accuracy of 0.5 mm for targets' height of 40 mm. This method enables first to clearly identify the position of the beam on the target and secondly to evaluate the process of pin-holes or surface cracks formation according to heavy ion beam dose. With this information, the heavy ion beam position and time structure on the targets can be corrected on-line, moreover it helps in the decision of replacing the targets due to material losses [25].

3.2 Monitoring with α–particle emission

3.2.1 Energy loss

For targets thin enough to slow down alpha particles emitted by a source, the measurement of the alpha energy loss is well suited to deduce the effective thickness of the target whose composition and stopping power are known [5], [26-28]. This technique can be implemented in vacuum reaction chambers of rotating targets. By placing the wheel between the source and the detector away from the heavy ion beam direction, an on-line evaluation (depending on the source activity) of the targets thickness as a function of their length is achievable with accuracies of ±5%.

3.2.2 α activity

For actinide targets, when the beam is stopped and the spectrometer is tuned to the α emission velocity, the α activity of targets is measured at the focal plane. As described in [29, 30], the α spectra measured before and time to time during the experiment show various energy distributions according to the heavy ion beam dose. The position and width of the peaks reveal the changes in the targets' structure at each step of the conditioning process. Indeed, the narrowing of the low energy tail (see figure 3 of [29]) indicates that volatile contaminants from the targets, originated from rests of solvents, resulting from the production process involving electrochemical deposition, evaporates. The saturation of the shape of the spectrum reveals that the targets' structure changes into a homogeneous transparent glass-like layer during irradiation.

As plotted in figure 6 of [30], a gradually growth of the low-energy part suggested that the target was covered by a layer of some material which was confirmed by its visual inspection.

In addition, the losses of material due to irradiation can be checked by two methods: firstly with an α source moved into the beam position in front of the target, where the counting rate of the α particles from the target is compared to the one from source; or secondly by analysing the charge equilibration foil which collects the scattered target material [29].

The α-spectra obtained at the focal plane during experiment reveals relatively large shift and broadening of the α lines compared to the ones measuring the scattered target material collected in the charge equilibration foil after experiment. The biggest contributions (estimated to be ≈ 60%) to these differences are due to the energy loss of the α particles in the materials (targets and detectors before the focal plane detector), nevertheless a further contribution arises to the mechanical change of the targets which become dented during irradiation [29].

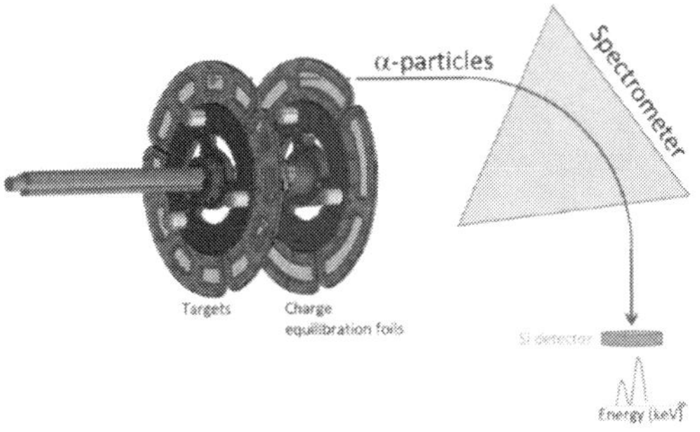

Fig. 3. Scheme of the a-activity method for active targets.

3.3 Rutherford scattering

One common technique to monitor the target thickness at the beam impact during irradiation is to register the elastically scattered projectiles and normalize their count rate to the beam current. The detection is performed with scintillation detectors mounted at opposite angles (± 30° in fusion-evaporation reaction) relative to the beam axis. The accuracy of the measurements depends on the precise knowledge of the solid angle of the detector.

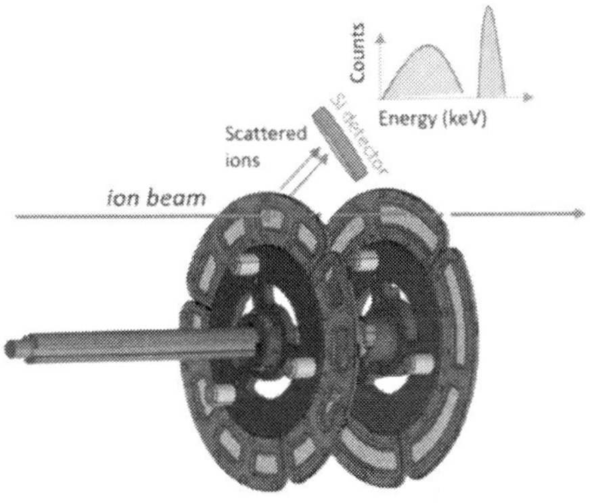

Fig. 4. Principle of the Rutherford scattering method to monitor targets' thickness.

3.4 Temperature

Due to the increase of beam intensities, targets' heating is a major issue. The target temperature can be monitored on-line by means of pyrometer [31] or infra-red devices (camera, thermo-viewer or fiberscope) [32-34]. As the targets are in a high-radiation environment, the thermal systems have to be shielded and often deported on large distance, which requires precise adjustment of the measuring position and efficient optical mirror and lenses. With the system of [33], a position resolution on targets of 0.5 mm is achieved. An essential parameter in temperature measurements is the overall emissivity on the target surface. It can be determined with heated calibrators and/or part of the surface to be analysed with known emissivity, e.g. heat-resistant black paint covering the material [34].

Complementary devices, such as CDD camera or endoscopes are used to monitor the beam spot position [33].

4 Summary

Due to the increasing demand for more accurate data and to the upgrade of accelerators, high quality and stable targets with precisely known properties are requested. As listed above, techniques are on common use to characterize actinide targets in order to improve the manufacture processes and to monitor and control them during irradiation in order to maintain their integrity. These techniques are also essential for the development and evaluation of new target production techniques.

For their help in achieving this review and for giving me materials for the oral presentation, I want to express my gratitude to my colleagues Goedele Sibbens, Klaus Eberhardt, Bettina Lommel and Anna Solarz from European commission, Joint Research Center, Geel; Universität Mainz, Institut fur kernchemie; GSI Helmholtzcentrum für Schwerionenforshung, Darmstadt and Heavy Ion Laboratory, University Warsaw respectively.

References

1. P. D'Hondt, C. Wagemans, A. Emsallem and R. Brissot, Ann. nucl. Energy **11, N₀ 10**, 485 (1984)
2. P. Schillebeeckx, A. Borella, J. C. Drohe, R. Eykens, S. Kopecky, C. Massimi, L. C. Mihailescu, A. Moens, M. Moxon and R. Wynants, Nuclear Instruments and Methods in Physics Research A **613**, 378 (2010)
3. D. Sapundjiev, A. Dean, U. Jacobsson, K. Luyckx, A. Moens, G. Sibbens, R. Eykens and Y. Aregbe, Nuclear Instruments and Methods in Physics Research A **686**, 75 (2012)
4. B. Lommel and B. Kindler, *Targets for Particle Accelerators in Encyclopedia of Applied High Energy and Particle Physics* (R. Stock, Éd., Weinheim, WILEY-VCH Verlag GmbH & Co KGaA, 2009)
5. A. Stolarz, Journal of Radioanalytical and Nuclear Chemistry **299**, 913 (2014)
6. A. Mitu, *this proceeding (*2019)
7. S. Pommé, Metrologia **52**, S73 (2015)
8. D. M. Gilliam and A. T. Yue, Journal of Radioanalytical and Nuclear Chemistry **299**, 1061 (2014)

9. G. Sibbens, A. Moens, R. Eykens, D. Vanleeuw, F. Kehoe, H. Kühn, R. Wynants, J. Heyse, A. Plompen, R. Jakopic, S. Richter and Y. Aregbe, Journal of Radioanalytical and Nuclear Chemistry **299**, 1093 (2014)
10. J. Heyse, M. Anastasiou, R. Eykens, A. Moens, A. J. M. Plompen, P. Shillebeckx, G. Sibbens, D. Vanleeuw and R. Wynants, Journal of Radioanalytical and Nuclear Chemistry **299**, 1055 (2014)
11. R. Haas, S. Lohse, C. E. Düllmann, K. Eberhardt, C. Mokry and J. Runke, Nuclear Instruments and Methods in Physics Research A **874**, 43 (2017)
12. D. Liebe, K. Eberhardt, W. Hartmann, T. Häger, A. Hübner, J. V. Kratz, B. Kindler, B. Lommel, P. Thörle, M. Schädel and J. Steiner, Nuclear Instruments and Methods in Physics Research A **590**, 145 (2008)
13. G. Sibbens, M. Ernstberger, T. Gouder, M. Marouli, A. Moens, A. Seibert, D. Vanleeuw, M. Vargas Zuniga, T. Wiss, M. Zampella and E. Zuleger, AIP Conf. Proc. **1962**, 030007-1 (2018)
14. D. Liebe, K. Eberhardt, W. Hartmann, A. Hübner, B. Kindler, J. V. Kratz, B. Lommel and J. Steiner, Annual Report, Institut für Kernchemie, University of Mainz, A5 (2005)
15. A. Vascon, S. Santi, A. A. Isse, A. Kühle, T. Reich, J. Drebert, K. Eberhardt and C. E. Düllmann, Nuclear Instruments and Methods in Physics Research A **714**, 163 (2013)
16. S. Richter and S. A. Goldberg, International Journal of Mass Spectrometry **229**, 181 (2003)
17. P. Robouch, G. Arana, M. Eguskiza, M. I. Maguregui, S. Pommé and C. Ingelbrecht, Nuclear Instruments and Methods in Physics Research A **480**,128 (2002)
18. K. Eberhardt and N. Trautmann, Technical Reports Series International Atomic Energy Agency **455**, 537 (2007)
19. K. Eberhardt, *Private Communication*
20. O. A. Wasson and R. A. Schrack, Nuclear Instruments and Methods in Physics Research A **282**, 194 (1989)
21. C.-O. B. Bacri, C. Baumier, J. Bourçois, L. Delbecq, D. Ledu, N. Pauwels, S. Picard, S. Renouf and C. Tanguy, Nuclear Instruments and Methods in Physics Research A **406**, 48 (2017)
22. C. O. Bacri, V. Petitbon-Thévenet, J. Mottier, H. Lefort, A. Durnez and F. Fortuna, Journal of Radioanalytical and Nuclear Chemistry **299**, 1099 (2014)
23. R. Mann, *Patent DE 10242962 A1 GSI*
24. R. Mann, D. Ackermann, S. Antalic, H.-G. Burkhardt, P. Cagarda, D. Gembalies-Datz, W. Hartmann, F. P. Hessberger, S. Hofmann, B. Kindler, P. Kuusiniemi, B. Lommel, S. Saro, H. G. Schött and J. Steiner, GSI Scientific Report **2003**, 224 (2004)
25. J. Kallunkathariyil, C. Stodel, C. Marry, G. Frémont, B. Bastin, J. Piot, E. Clément, S. Le Moal, V. Morel, J.-C. Thomas, O. Kamalou, C. Spitaëls, H. Savajols, M. Vostinar, F. Pellemoine and W. Mittig, AIP Conf. Proc. **1962**, 030019 (2018)
26. R. Bimbot, S. Della Negra, C. Deprun, D. Gardes and M.-F. Rivet, IPNO RC **79 06** (1979)
27. H. L. Adair, Nuclear Instruments and Methods **102**, 599-610 (1972)
28. «AASI - Program for simulating energy spectra in alpha spectrometry» Available: https://www.stuk.fi/web/en/services/aasi-program-for-simulating-energy-spectra-in-alpha-spectrometry
29. S. Hofmann, S. Heinz, R. Mann, J. Maurer, J. Khuyagbaatar, D. Ackermann, S. Antalic, W. Barth, M. Block, H. G. Burkhard, V. F. Comas, L. Dahl, K. Eberhardt, J. Gostic, R. A. Henderson, J. A. Heredia, F. P. Hessberger, J. M. Kenneally, B. Kindler, I. Kojouharov, J. V. Kratz, R. Lang, M. Leino, B. Lommel, K. J. Moody, G. Münzenberg, S. L. Nelson, K. Nishio, A. G. Popeko, J. Runke, S. Saro, D. A. Shaughnessy, M. A. Stoyer, P. Thörle-Pospiech, K. Tinshert, N. Trautmann, J. Uusitalo, P. A. Wilk and A. V. Yeremin, European Physical Journal A **48**, 62 (2012)
30. N. T. Brewer, V. K. Utyonkov, K. P. Rykaczewski, Y. T. Oganessian, F. S. Abdullin, R. A. Boll, D. J. Jean, S. N. Dmitriev, J. G. Ezold, L. K. Felker, R. K. Grzywacz, M. G. Itkis, N. D. Kovrizhnykh, D. C. McInturff, K. Miernik, G. D. Owen, A. N. Polyakov, A. G. Popeko, J. B. Roberto, A. V. Sabel'nikov, R. N. Sagaidak, I. V. Shirokovsky, M. V. Shumeiko, N. J. Sims, E. H. Smith, V. G. Subbotin, A. M. Sukhov, A. I. Svririkhin, Y. S. Tsyganov, S. M. Van Cleve, A. A. Voinov, A. A. Vostokin, C. S. White, J. H. Hamilton and M. A. Stoyer, Physical Review C **98**, 024317 (2018)
31. E. Jäger, H. Brand, C. E. Düllmann, J. Khuyagbaatar, J. Krier, M. Schädel, T. Torres and A. Yakushev, Journal of Radioanalytical and Nuclear Chemistry **299**, 1073 (2014)
32. S. Antalic, P. Cagarda, D. Ackermann, H.-G. Burkhardt, F.-P. Hessberger, S. Hofmann, B. Kindler, J. Kojouharova, B. Lommel, R. Mann, S. Saro and H.-J. Schött, Nuclear Instruments and Methods in Physics Research A **530**, 185 (2004)
33. A. Yoshida, T. Suda, T. Ohtsuki, H. Yuki and T. Kubo, Nuclear Instruments and Methods in Physics Research A **590**, 204 (2008)
34. A. Yoshida, Y. Yanagisawa and T. Kubo, Nuclear Instruments and Methods in Physics Research A **655**, 10 (2011)

Surface and thickness measurement in the Targetlab of GSI

Birgit Kindler*, *Elif* Celik Ayik, *Annett* Hübner, *Bettina* Lommel, *Jutta* Steiner, and *Vera* Yakusheva

GSI Helmholtzzentrum für Schwerionenforschung, Targetlabor, Planckstrasse 1,64291 Darmstadt, Germany

Abstract. For characterization of targets and foils prepared at the target laboratory as well as for characterization of e.g. degrader or windows of internal customers, different analytical devices are available. Besides a lot of standard equipment, the target laboratory of GSI holds a 3D-measurement system (MicroProf®) equipped with optical sensors for measuring surface parameters as well as total thickness variations contact-free. In the paper the measuring principle including the possibilities and features of the MicroProf®-system are explained and some different applications are shown.

1 Introduction

After a short introduction to the tasks of analytics in the target laboratory at GSI and an overview of the analytical tools and methods available, we will concentrate on the MicroProf®, a machine purchased from FRT GmbH. We will describe the setup, the measuring principle and show its possibilities with the help of some examples of application.

1.1 Analytics at GSI target laboratory

Analytics plays a major role in the GSI target laboratory. First, we have to determine the key parameters of our targets and foils, sometimes even of our starting materials, as good as possible. Key parameters are for example weight, thickness, surface features like topology or roughness, composition and impurities.

We need different analytical methods for quality control of the samples produced in our lab as well as for parts produced externally. For most of the samples leaving our laboratory, detailed documentation of the sample parameters is stored.

1.2 Analytical tools available

The working horses in the target laboratory are a number of balances from Mettler Toledo: a microbalance with a readability of 0.1 µg, several analytical balances with a readability range from 0.01 to 0.1 mg, and one high-capacity balance, suitable to measure bulky and heavy samples up to several kg.

Incremental length gauges from Heidenhain with a travel of 10 mm and 60 mm, respectively, are applied for a quick determination of the thickness of foils and sheets.

For measuring the thickness of thin, transparent films, a double-beam UV-vis spectrophotometer from GBC is used. It works in the wavelength range from 90 to 190 nm and is a fast method for materials for which we worked out a calibration curve. Especially for thin carbon and gold foils, this method is irreplaceable for us.

For investigating the topology and sample surface, inspecting and documenting the quality of coatings, and for checking foils for pinholes two microscopes are available: A stereo-microscope from Leica with a zoom magnification from 6.3x up to 50x and an Orthoplan Largefield microscope from Leitz with an objective revolver with magnification from 24.6x to 640x with incident as well as with transmitted light. Both microscopes are equipped with a digital camera and with and image acquisition software.

For investigating the morphology and topology of small samples, a Scanning Electron Microscope from Philips is available, which is equipped with a detector for energy-dispersive x-ray analysis from AMETEK This is is a powerful tool for determining the composition of samples, for investigating impurities and film thicknesses.

The Target Scanner [1] is a self-constructed machine for measuring the rough topology and the thickness of bulk samples with dimensions up to 100 cm times 60 cm with an absolute accuracy of thickness of about 1 µm. This machine works with two vertically aligned incremental gauges touching the surfaces of the sample from both sides. This method works well but has the disadvantage that the gauges have to work with a certain contact pressure in order to get reliable measuring values what makes the investigation of ductile or sensible surfaces impossible.

Therefore, we purchased in 2011 the MicroProf®-System from FRT GmbH [2], which has a comparable arrangement of components but works with touch-free optical sensors.

* Corresponding author: b.kindler@gsi.de

2 Features of the MicroProf®-system

2.1. Hardware

The MicroProf® system, as shown in Fig. 1, consists of a tabletop and a yoke made of massive granite, which is mounted on a pneumatic-mechanical system ensuring a decoupling of mechanical disturbance from the floor. On the granite tabletop, a XY-stage is attached with a travel of 420 mm and 310 mm, respectively, thus defining the maximum dimensions of a sample that can be scanned during a measurement. Different fixtures can be mounted on the table, appropriate for holding the particular sample and the calibration measures needed for the determination of the total thickness variation (TTV) of a sample.

Fig. 1. MicroProf®-system.

The sensor head is situated on a vertical slide with a travel of 160 mm thus defining the maximum height of samples to be measured.

On the sensor head, multiple devices can be mounted in parallel in order to measure the same place on a sample with different methods, the lateral offset of the devices being automatically corrected via the controlling software. In our current setup, a CCD camera for sample positioning and two chromatic point sensors are mounted on the sensor head.

We have one sensor with a measuring range for 25 mm and a working distance of 80 mm, which allows for measuring large steps and very wavy samples with a lateral resolution of 14 µm and a vertical resolution of 8 µm.

Furthermore, two sensors with a measuring range of 600 µm and a working distance of 6.5 mm are available. They have a lateral resolution of 1 – 2 µm and a vertical resolution of 200 nm. They are mounted vertically aligned in a so-called TTV arrangement, the second sensor being fixed upside down to the granite table plate thus "looking" to the sample from the bottom side. For a TTV measurement, the system has to be calibrated at first with a standard measure that has a thickness in the range of ± 0.6 mm to that of the sample. After the calibration the thickness variation and the topology of the top and bottom surfaces of the sample can be measured simultaneously.

2.2 Software

The *Acquire* software controls the FRT metrology tools and all sensors. This software allows fast and easy manual measurements in 1D (line) or 2D (area) and gives a quick impression of the measurement data.

FRT Mark III is the analysis software for the evaluation of surface and profile measurements. It allows the easy and efficient evaluation of surface structures and parameters according to common international standards (DIN EN ISO, SEMI, etc.). Versatile 3D views represent the surface data in realistic detail.

2.3 Measuring principle

The measuring principle is depicted in Fig. 2. An optical fibre transfers white light from a halogen lamp to the small and passive sensor head. The optical system in the head contains an aspheric lens with strong chromatic aberration that spreads the different wavelengths of the white light in vertical direction so that the focus for each wavelength is situated at a different distance from the end face of the fibre. Therefore, the confocal imaging of the sample surface is fulfilled for only one wavelength that is reflected back to the optical system and is then guided via the optical fibre coupler to the spectrograph. In the spectrograph, the signal height at the peak wavelength is then evaluated and translated to a topology signal.

By moving the sample on the xy-stage relative to the stationary sensor head, the sample topology or, for TTV arrangement, the sample thickness can be recorded.

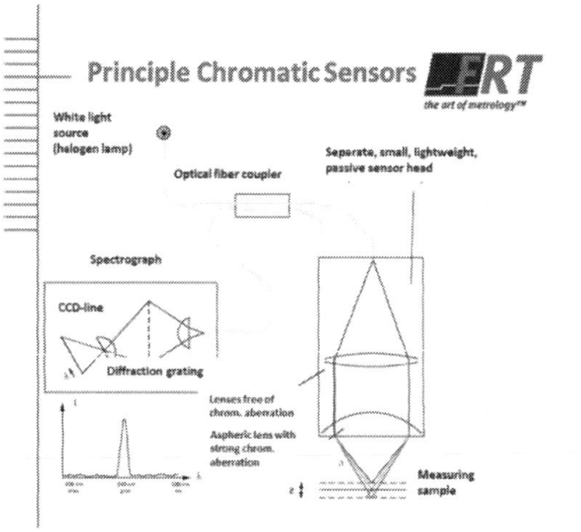

Fig. 2. Measuring principle of chromatic sensors.

3 Examples of application

To demonstrate the versatility of the system some examples for measurements with the MicroProf® will be shown in the following paragraphs.

3.1 Test step

To get a first impression, two standard stainless steel measures with a thickness of 1.01mm and 1.005 mm, respectively, are placed side by side such creating a step with a height of 5 µm. In Fig. 3 the thickness data are shown in a 3-dimensional plot.

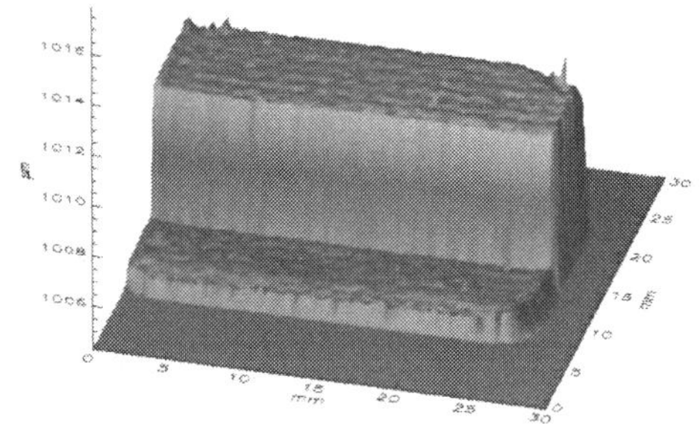

Fig. 3. 3D-plot of a 5 µm step.

In Fig. 4a and c the same measurement data as in Fig. 3 are presented in a 2-dimensional plot with the thickness now being colour-coded. The thickness values along the green line in 4a and the red line in 4c are plotted in the profile graphs in the pictures 4b and 4d on the right. The profile graph in 4b along the green line runs parallel to the plane of the 1.01 mm standard measure demonstrating the vertical accuracy of the system. The profile graph in 4d runs across the step thus showing the accuracy of mapping the sharp step.

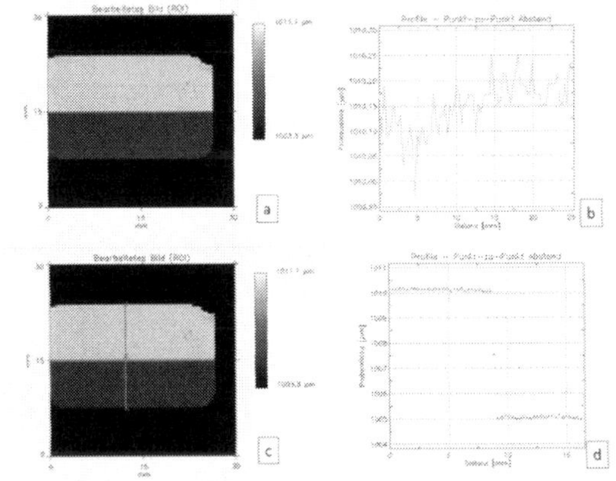

Fig. 4. 2D colour-coded thickness data of the test step with profile lines parallel to the plane in 4a and across the step in 4c. 4b and 4d show the thickness data along the profile lines.

3.2 Monitoring preparation steps

The TTV-measurement is a mighty tool to monitor the surface structure and the sample shape after each preparation step. In Fig. 5 the thickness measurement of a carbon sample is shown as a colour-coded 2-dimensional plot after first lapping the sample (5a) and then after polishing (5c) with the corresponding graphs along the profile lines (5b and 5d respectively).

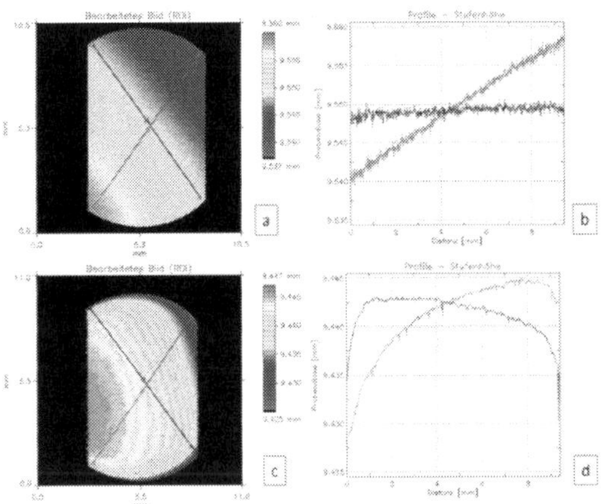

Fig. 5. Thickness of a carbon sample after lapping (top) and after polishing (bottom) 2D colour-coded (a and c) and in a graph along the profile lines (b and d).

It is clearly visible that the surface structure after lapping (Fig. 5a and 5b) is microscopically very rough, that the sample surfaces are very flat but that the whole sample is a bit wedge-shaped (Δ ~ 15 µm along the red profile line). After polishing (Fig. 5c and 5d) the surface structure is much smoother but the wedge shape remains. The sample shape is more convex and globular, the edges of the sample being thinner than the centre.

This is essential information for the application of the target since the user has to decide if the homogeneity of the sample is more important or the smoothness of the surface(s). Visualizing such effects is very instructive for the users.

3.3 Quality control for purchased targets

Some materials like e.g. beryllium cannot be produced or reshaped in-house because of safety restrictions. Especially for those samples, it is important to have a reliable characterization since the specification of the producer is often not very significant or even incorrect.

In the example in Fig.6 two beryllium samples with a diameter of 20 mm and a thickness of 1.0 mm and 1.46 mm, respectively are evaluated. Though the nominal thickness is pretty much comparable and both samples were purchased from the same source, the features of both samples are already quite different when only comparing the appearance (photorealistic view of thickness values pictures in the first column of Fig.6). The pictures of the 2-dimensional colour-coded thickness as well as the graphs of the corresponding profile lines clearly show that the sample with 1 mm nominal thickness (Fig. 6 bottom row) is much more homogeneous in thickness compared to that of the 1.46 mm sample (Fig.6 top row) which is about 5% thinner in the centre compared to the edge.

This difference in quality can also be seen in the statistical evaluation in the last column of Fig. 6, which can be performed automatically with the Mark III software.

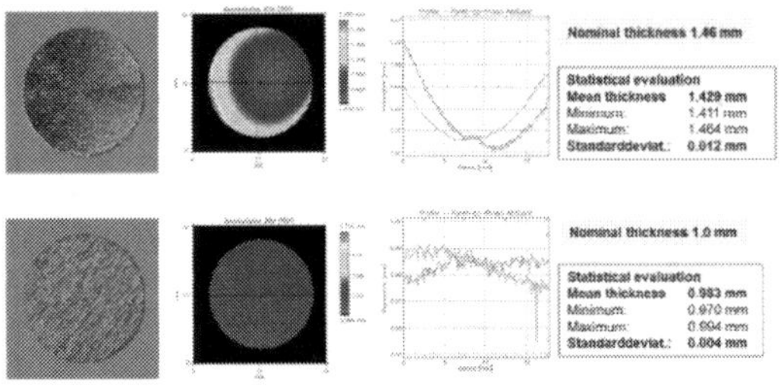

Fig. 6. Thickness evaluation of two beryllium samples with a nominal thickness of 1.46 mm (top) and 1.0 mm (bottom). From left to right: Thickness values as photorealistic view, 2D colour-coded, as graph along the profile lines, and the statistical evaluation (with the standard deviation from the mean).

3.4 Quality control of degraders

A special challenge is the measuring of large parts that are not flat but wedge-shaped like the degrader system shown schematically in Fig. 7. Since the sensors have a measuring range of only 0.6 mm, a slope of more than that cannot be measured at once.

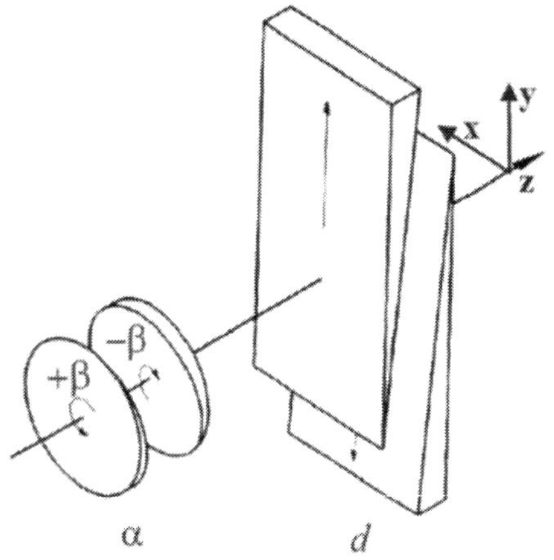

Fig. 7. Schematic drawing of the degrader system of the Fragment Separator [3].

In the following, the measurement of one of the wedged-shaped disks with a diameter of 330 mm and a thickness of 0.5 mm to 3.2 mm from one end to the other is shown in Fig. 8 on the left. The disk is mounted on the specimen holder with the flat side down, since the bottom sensor of the setup is fixed in position, and such that the slope on the upper side of the disk is increasing from left to right. Since the disk has no stop angle, the disk is at first placed by visual judgement and then the alignment is refined by measuring the thickness parallel to the edge where the thickness should be constant for an ideal alignment of the disk. If a variation in the thickness is measured the disk position is corrected manually, then the thickness is measured again and so on until a satisfactory alignment is obtained.

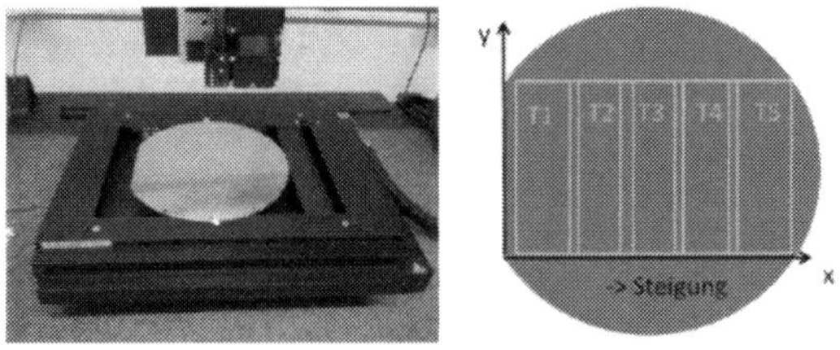

Fig. 8. A degrader disk mounted on the MicroProf®-table (left) and the schematic drawing of the disk with the overlapping measurement areas T1 – T5.

Since the total thickness difference across the disk is 2.7 mm, the measurement is subdivided into 5 overlapping parts, as indicated in Fig. 8 on the right. For each measurement part, the upper sensor has to be shifted upwards and a new calibration with an adequate standard measure has to be conducted.

The data obtained are then exported in ASCII-format and can be further evaluated with standard codes. Fig. 9 shows the colour-coded data of the 5 partial measurements, assembled and fitted for an overview of the quality of the whole part. As a result, the accuracy of the part was sufficient for the operation in the degrader system.

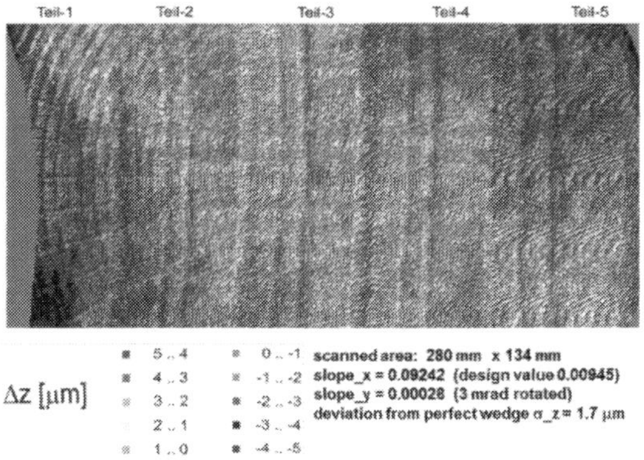

Fig. 9. Assembled and fitted data of thickness [4] of the degrader disk coming from the overlapping measurement areas measurements T1 – T5 as shown in Fig.8.

4 Summary and perspectives

The described applications showed some of the powerful features of the Microprof®-system we already experienced. Additionally to the examples presented, we also measured roughness parameters; we investigated thin and partly destroyed foils. By moving one sensor 2 µm aside also the thickness of transparent materials like glass or polypropylene have been measured.

https://doi.org/10.1051/epjconf/202022902002

In the future, we want to explore the system further to expand the measuring and evaluation possibilities. As a long-term perspective, we could consider extending our system with an additional sensor like for example an Atomic Force Microscope or a Film Thickness sensor.

References

1. B. Kindler, H. Brand, W. Hartmann, J. Steiner, J. Klemm and B. Lommel; Nucl. Instr. and Meth. **A 480**,160-165 (2002)
2. https://frtmetrology.com/en/
3. H. Geissel et al., Nucl. Instrum. Meth. B **70**, 286 (1992)
4. Credits to Dr. Helmut Weick, Fragment Separator Group of GSI.

Manufacturing and characterization of targets at IFIN-HH: developing an interdisciplinary body of knowledge

Andreea Mitu[1,2,*], *Nicoleta* Florea[1], *Nicolae* Mărginean[1], *Raluca* Mărginean[1], and *Gheorghe* Căta –Danil[2]

[1]Nuclear Physics Department, "Horia Hulubei" National Institute for Physics and Nuclear Engineering (IFIN-HH), Romania
[2]Physics Department, University POLITEHNICA of Bucharest, Romania

Abstract. Target manufacturing is one fundamental issue in nuclear physics experiments using accelerators. A variety of targets are required, each having to satisfy specific conditions related to the experimental particularities. In this context, a brief description of the target preparation laboratory developed at IFIN-HH is presented in this paper. To fulfill the specific requirements, the laboratory is endowed with high performance equipment for evaporation-condensation (thermal resistance, e-based systems, sputtering) and cold rolling. During the last years, consistent technological improvements were achieved. Target characteristics, quality and reliability are important for our experiments' feasibility in the first place, but also for the degree of confidence in the assumed accuracy. Consequently, XRD, AFM, SEM/EDX and RBS analyses are performed in collaboration with specialized departments from our institute and from other research centers.This paper is meant to synthetize our work so far.

Introduction

The target laboratory in IFIN-HH (National Institute for Physics and Nuclear Engineering – "Horia Hulubei") was developed to respond the experimental requirements with thin films of enriched stable isotopes ranging from tens of nanometers to a few micrometers. It was founded in 2013 [1] in order to fabricate targets mainly for nuclear physics experiments at both the 9 MV Tandem and the 3 MV Tandetron accelerators at IFIN-HH, but also for other research facilities (e.g. CERN, IPN Orsay, LNL). In order to improve the present capabilities of the laboratory, a part of the equipment was upgraded (See Section 1 – *Laboratory upgrade).*

Moreover, for the manufacturing of reliable targets from materials with extremely high vaporization temperatures in the range of 4500-5500°C (e.g. Os, W), the Pulsed Laser Deposition (PLD) technique was developed in collaboration with Photonic Processing of Advanced Materials laboratory (PPAM) from National Institute for Laser, Plasma and Radiation Physics (INFLPR). Also, in the same laboratory, we successfully fulfill thermal decontamination using a hydrogen oven in order to remove the oxygen from the surface of the target, which can give significant interference in the gamma spectra for some types of experiments.

As already mentioned, the final target characteristics strongly affect the precision of data measurements. According to the paper reported by Anna Stolarz, the most important parameters for a completed target are thickness uniformity and impurity identification [2]. In this context, the obtained targets are fully characterized using multiple advanced techniques such as Rutherford Backscattering Spectrometry (RBS), X-ray Diffraction (XRD), Atomic Force Microscopy (AFM), Scanning Electron Microscopy and Energy Dispersive X-ray Spectroscopy (SEM/EDX), which were performed in collaboration with specialized departments from our institute and from other research centers.

In 2016, IFIN-HH has managed to access €2 million funds through the European Union's "Competitiveness Operational Programme, 2014–2020" for the procurement of high tech equipment to establish the Romania's first national nuclear forensics laboratory.
The implementation of the project will provide the technical capabilities for handling and analyzing of different types of materials (radioactive and non-radioactive) including MO (optical microscopy), SEM/EDX and XRD.
Detailed and specific information about all the aforementioned techniques is given and finally the plan for our process and performance in the new configuration is presented in this paper. It appears the manufacturing process combined with the characterization techniques offer a strong perspective for target development methods.

1 Laboratory upgrade

The target preparation laboratory was developed to produce self-supported targets or deposited layers on various backings applying high vacuum evaporation or rolling techniques, dedicated to different nuclear physics experiments. As reported [1], the laboratory has two High Vacuum Deposition Systems (TE18-High Vacuum

Deposition System and Quorum Technologies E6700 Bench Top Evaporator) equipped with thermal evaporation, e-gun and sputtering technologies.

Recently, a second e-beam system produced by Mantis Deposition Systems was integrated in the vacuum deposition chamber of TE18-High Vacuum Deposition System (see Fig. 1). This system has numerous advantages for isotopically enriched film deposition from different materials with thicknesses ranging from a fraction of a monolayer to several hundreds of nanometers. Moreover, according to the studies reported by E.H. Kobisk, J.P. Greene and J.M. Heagney [3-5], it was proven to be very useful for the reduction of oxides having high reduction temperatures (1850-1950°C) in order to obtain a metallic form of the deposited film. The reduction reaction involves an electron heated crucible and zirconium as reducing agent.

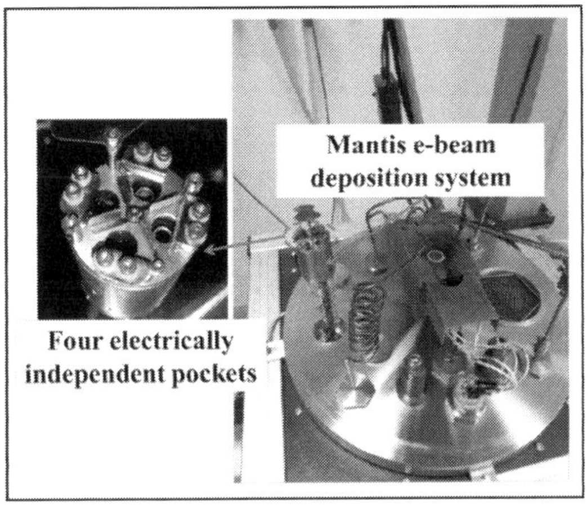

Fig. 1. TE18-High Vacuum Deposition System produced by Intercovamex Company upgraded with e-beam deposition system produced by Mantis Deposition

The Mantis e-beam system is equipped with a QUAD-EV-C source with a power supply of 500 W that can be used to heat the target material to temperatures up to 3000°C. The target rod/crucible is located close to the W filament. When the current through the filament is sufficient, the electron emission temperature is reached and a stream of electrons is drawn from the filament towards the rod/crucible. Their impact on the rod increases the temperature, which leads to evaporation of the rod or the material placed in the crucible. A Quad-EV-C source is used to ensure optimum control of the evaporation rate at low fluxes. This system is equipped with four pockets of 390 mm^3; each one uses independent filaments, flux and high voltage channels. It can be used for multilayer deposition by independent monitoring of the deposition rate. The main advantage of this system is that all parameters can be monitored and controlled by means of software applications [6].

At the commissioning date, the target laboratory also included an electrically controlled Durston type DRM 100 rolling mill. The technique of rolling is extremely efficient for self-supported metallic foil production using small amounts of starting material. It allows preparation of metallic foils with areal densities ranging from several hundreds of μg/cm^2 to hundreds of mg/cm^2. To increase the quality of the produced metallic foils by rolling, a more powerful, double sided, electrically controlled Durston type FSM 160 VCC rolling mill with variable speed (5-20 rpm) and with touch screen display was acquired (see Fig. 2) [7].

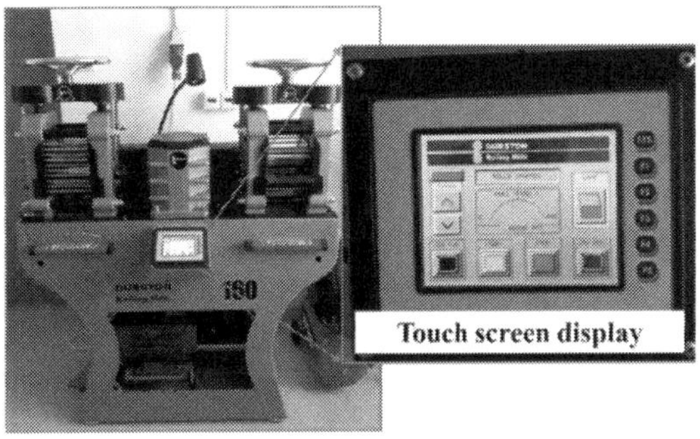

Fig. 2. Durston type FSM 160 VCC power rolling mill, floor standing, double sided with touch screen display

Another important upgrade for our laboratory consists in the procurement of an Atlas™ Automatic 25 Ton (25T) Hydraulic Press equipped with 40 mm, 32 mm, 20 mm, 13 mm, 10 mm and 5 mm evacuable pellet dies, produced by Specac (see Fig. 3) [8]. All the parts of the die are manufactured from hardened stainless steel for optimum quality and durability, and highly polished surfaces for contact with the sample. The applied pressure and release are well controlled. The LCD display shows press status and load conditions providing a digital display of the load applied. The hydraulic press is mainly used for producing self-supported pellets, which can be further used in experiments as thick targets.

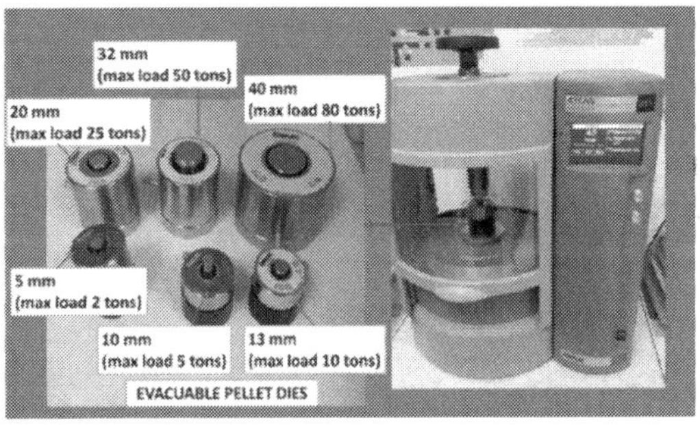

Fig. 3. Atlas™ Automatic 25Ton (25T) Hydraulic Presses (right) equipped with 40 mm, 32 mm, 20 mm, 13 mm, 20 mm and 5 mm evacuable pellet dies (left)

The weighing method is very important in the determination of the areal density of foils or deposited targets, combined with the precise surface measuring using a digital optical microscope (Dino-Lite Pro Digital Microscope AM-413TL). Thus, for more stable and accurate weighing results, the new XP56/M analytical balance from Mettler Toledo (Fig. 4) was purchased, offering a weighing capacity of 52 g with a 0.001 mg readability.

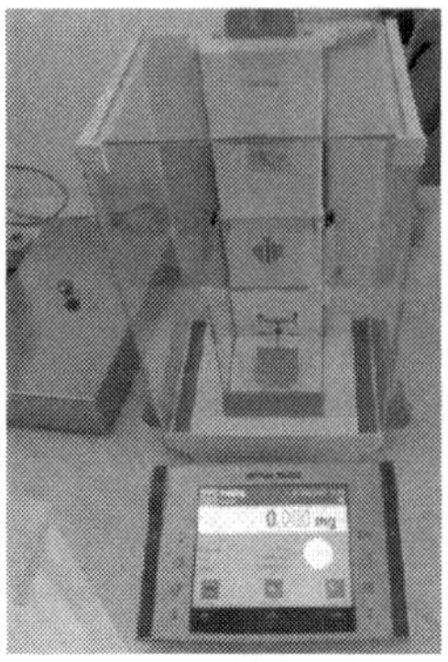

Fig. 4. XP56/M analytical balance

2 Specific target preparation

During the last years we prepared a large variety of targets using different methods, either available in the target preparation laboratory, or in collaboration with other laboratories as PPAM from INFLPR, that have complementary equipment (e.g. PLD method). For some specific experiments that required special characteristics, the targets were characterized by means of RBS, AFM, SEM/EDX and XRD techniques. Table 1 shows examples of targets prepared in IFIN-HH.

Table 1: Examples of produced targets by rolling, redox, pulsed lased deposition (PLD) and thermal evaporation (TE)

Element	Backing	Areal density	Method	Comment
^{64}Ni	no backing	1 mg/cm^2 5 mg/cm^2	Rolling	a)
^{24}Mg	no backing	0.2 – 5 mg/cm^2	Redox/ rolling	b)
W	Al	1 μm / 20 μm	PLD	c)
Os	Al	100 nm/ 20 μm 2 μm/ 20 μm	PLD	d)
^{196}Pt	no backing	9 mg/cm^2	rolling	e)
^{106}Cd	Ta	0.5 mg/cm^2/ 2mg/cm^2	rolling	f)
SrF_2	Al	1.6 mg/cm^2	TE	g)
$SrCO_3$	Al	2.4 mg/cm^2	TE	h)
^{57}Fe	no backing	0.4 mg/cm^2	rolling	i)
^{96}Zr	no backing	1 mg/cm^2 2 mg/cm^2	rolling	j)
^{65}Cu	no backing	7 mg/cm^2	rolling	k)
^{64}Ni	Fe	0.7 mg/cm^2/ 5 mg/cm^2	TE	l)

a) The ^{64}Ni as metallic powder was pressed in a pellet, heated with an e-gun and finally rolled until the desired thickness was obtained;
b) Various ^{24}Mg foils with different thicknesses were prepared starting from ^{24}Mg oxide powder, which was reduced to a metallic form using fresh lanthanum by evaporation-condensation. The material was placed in a Ta boat covered with a copper cooled device;
c) Natural W targets were manufactured by PLD method, Nd:YAG 1064 nm;
d) Natural Os targets on Al backing were obtained by PLD, Nd:YAG 1064 nm;
e) The powder was sintered in a solid piece using the e-gun system and the standard rolling procedure was applied;
f) A target of ^{106}Cd deposited on a Ta backing was obtained by rolling both foils together for a plunger type interaction chamber. With a slight pressure between the rolls, the Cd will sticked to the Ta and could be rolled down together, obtaining a thickness reduced by a factor of four;
g) A homemade cylindrical Ta boat was used for the evaporation process and no difficulties occurred;
h) A similar Ta boat to the previous one (g) was used. Because of the high vaporization temperature of the carbonate an overheating of the copper electrodes can cause contamination of the target;
i) It could be rolled below the limit value estimated in the literature [9], which is 0.5mg/cm^2;
j) The rolling of ^{96}Zr is very difficult. In order to prepare plunger targets with a large surface, special care must be attributed to avoid cracks and holes - a thermal treatment is recommended after every 3 microns thicknesses decrease,
k) ^{65}Cu foils were obtained by rolling. To improve the measurements for the proposed experiment, which was a transfer reaction, a thermal treatment was applied for oxygen decontamination in order to obtain clear γ-spectra [23];
l) The adhesion between Ni and Fe is very poor; to enhance the adhesion between these two materials, the Fe surface used as backing was scratched with a Ted Pella, Pelco Bell Jar Kleen polishing cleaner powder and then washed, thus the surface roughness allowed adhesion of the Ni layer.

3 Target characterization

A wide range of nuclear physics experiments requires high-quality and well-characterized targets. In order to fulfill such demands, various techniques were applied. The target thickness is one of the most important parameters, having a considerable contribution in all types of experiments. One of the most common techniques that we use for thickness determination is the α-transmission technique using an α source, which contains a mixture of ^{241}Am and ^{244}Cm [1]. Moreover, the thicknesses of different targets were also measured by the RBS technique using α or proton beams delivered by the 3 MV Tandetron accelerator available at IFIN-HH. The areal density of the target is very important in all target-beam experiments, as the cross-section uncertainties scale directly with the areal density error. The backscattered protons were registered with an AMETEK type BU-012-050-100 charged particle detector; the RBS spectra were simulated using the

SIMNRA software package [10-12]. Therefore, in order to investigate the phase composition, the crystalline structure and the chemical purity of the targets, XRD patterns were registered with a PANalytical X'Pert MRD [13], instrument equipped with a fixed divergence slit and fitted with a pixel detector. The X-rays were generated by a Cu anode polarized at a 40 kV potential, which was bombarded by an electron current of 30 mA. The diffraction data were acquired in a reflection set-up.

AFM analyses were performed with a Park System XE-100 microscope. The studies were conducted in the non-contact mode using silicon cantilevers. The AFM images show topographic information with 256×256 pixel resolution.

The surface morphology of the targets was investigated by performing SEM/EDX measurements using two types of electron microscopes: the FEI Inspect S microscope equipped with a Si(Li) detector EDAX, and the Zeiss EVO MA15 microscope coupled with the Thermo Scientific EDX system.

4 Specific examples and particularities

The target preparation laboratory at IFIN-HH in collaboration with INFLPR was involved in the preparation of some particular targets used in different experiments. The target material occasionally exhibits difficulties such as high vaporization temperatures (osmium), high reactivity in air (strontium) or oxygen absorption on the material surface (nickel). Further, a brief description about the preparation and characterization of Os, Sr compounds and ^{64}Ni targets will be reported.

4.1 Osmium targets

The goal of this research was to prepare Os targets on Al backings as thin films using the PLD technique. Osmium is the densest stable element; the vaporization temperature is high (~5000°C), which makes evaporation a challenging process. Two different Os targets with an areal density of 0.2 and 5 mg/cm^2 were successfully deposited onto Al backings by the PLD technique using an Nd:YAG laser with a wavelength of 1064 nm (Fig. 5) [14-17].

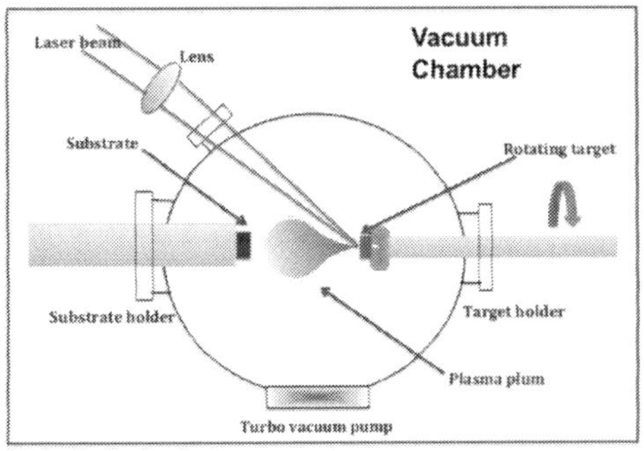

Fig. 5. Experimental set-up for PLD method [17]

Characteristics of the resulting thin films are used in accelerator-based experiments for alpha particle induced reactions. The Os targets were characterized using various techniques as follows. The thicknesses of the Os films were determined by means of RBS, using a 1.8 MeV proton beam. The morphological features of the obtained targets analyzed by AFM and SEM/EDX reveals that the deposited layers (0.2 and 5 mg/cm^2) follow the substrate topography and exhibits good uniformity, meaning it was equally distributed on the whole surface according to the mapping micrographs, while the aluminum backing exhibits imperfections. The EDX and XRD spectra confirm the target purity, as the peaks from the deposited layers are perfectly correlated with the peaks from both the bulk material and the Al backing.

The acquired data during the nuclear activation experiment is presently analyzed in order to establish if all reactions of interest took place. Extended information on this topic and its particularities is available in a research paper [17].

4.2 Targets from strontium compounds

Based on previously reported results [18], different SrF_2 and $SrCO_3$ targets were prepared by thermal evaporation in similar conditions for nuclear astrophysical experiments with alpha particles accelerated near the Gamow window energy [19, 20]. For this type of experiments, the metallic Sr targets cannot be used, since they exhibit a significant chemical activity and oxidize very fast in air. In this context, the target characteristics (thickness, chemical purity and surface morphology) for both Sr compounds were determined by α transmission measurements, XRD and SEM/EDX analysis, in order to establish which one is better suited for this experiment. The results reveal that SrF_2 target exhibits better properties compared to the $SrCO_3$ one. This result was supported by the γ-ray spectra collected in identical conditions of beam fluency corresponding to the two compounds, showing a clear difference for the intensity of the gamma rays of interest. Partial results of the cross-section measurements are reported in [21]. Detailed information about the strontium compounds target production and characterization, detailed results and discussion can be found in the extended research paper already published [18].

4.3 Decontamination of ^{64}Ni targets

According to [22], high purity ^{64}Ni targets were prepared for two neutron transfer reactions of an ^{18}O beam on ^{64}Ni targets at sub-barrier energies [23]. Target preparation considers the Ni chemical activity in air, as oxygen layers are incorporated and formed on the target surface. This is a major drawback, since gamma rays from ^{16}O target reactions bring a significant contribution to the recorded spectrum. In order to avoid the contamination of the γ spectra, a thermal treatment with a hydrogen oven was applied [24]. This way we obtained a significant reduction of the O contamination (confirmed

 https://doi.org/10.1051/epjconf/202022903001

by the gamma lines in the spectra as shown in Fig. 6). Detailed information about the treatment procedure and its effect can be found in our research paper [22].

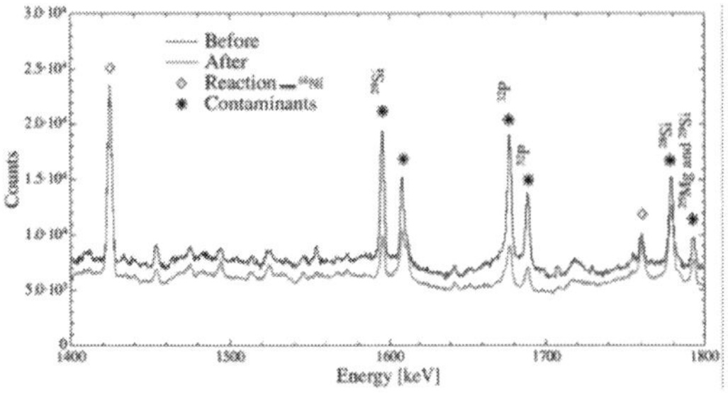

Fig. 6. Gamma spectrum for the reaction of interest before/after thermal treatment of the target [22]

5. Developing an interdisciplinary collaboration

IFIN-HH has managed in 2016 to access funds through the European Union's "Competitiveness Operational Programme, 2014–2020" for the procurement of high tech equipment to establish the first Romanian Nuclear Forensics laboratory. The facility will be used for the analysis of forensic materials that contain uranium, plutonium and their descents (see Fig.7). This objective was defined in order to consolidate the national security and research capabilities in forensics.

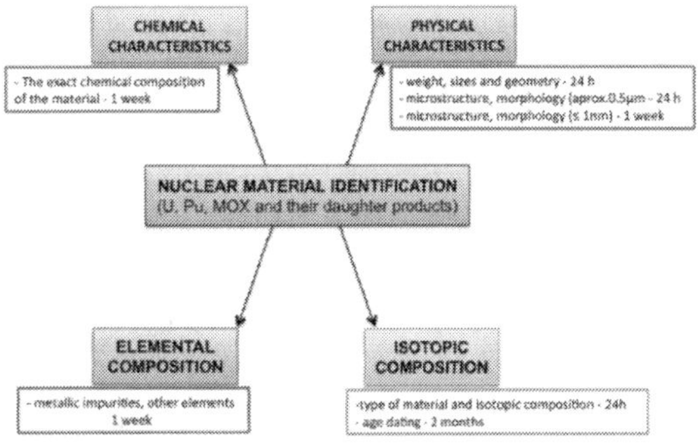

Fig. 7. Overview of analysis techniques required for the categorization and characterization of nuclear materials

Analyses start with physical characteristics: the sample weight is determined with an analytical balance; the sample size and shape are determined by classical methods and a 3D scanner. This information can lead to the first clues on origin (place of extraction or production), processing procedure, conditions for storing or transport, etc.

For the macroscopic parameters, an inverted optical microscope with examination in polarized light capabilities is used. A high-resolution electronic microscope (≤1 nm for secondary electrons imaging), equipped with an energy dispersive X-ray detector will be configured in order to allow characterization of a wide range of materials. Then, XRD is used to characterize phase composition and the crystalline structure of the sample.

Then the chemical composition is considered. The high concentrations of the elements define the material, while minor concentration (impurities) can lead to identification of the material and help determining its origin.

Ion Beam Analysis (IBA) techniques are based on atomic or nuclear interactions of accelerated particle beams (in MeV energy magnitudes) with atoms or nuclei of an irradiated sample. The Particle Induced X-ray Emission (PIXE) technique is based on the detection of X-rays, which are emitted after the interaction of a beam (in general, relatively low energy protons) with atoms from the sample, allowing qualitative and quantitative analysis. The Particle Induced Gamma-Ray Emission (PIGE) technique is based on the capture of the accelerated particle by the nucleus from the sample, leading to a compound, newly formed nucleus (short life time); followed by emission of protons, alpha particles, and neutrons or/and prompt gamma rays.

Also, Accelerator Mass Spectrometry (AMS) is the most sensitive elemental analysis method available. Isotopic ratios can reach 10^{-16} isotope/element; therefore, it is used for measuring low levels of long-lived radionuclides and trace elements.

Gamma-Ray Spectrometry, a non-destructive technique, is used to identify gamma-emitting radioactive nuclides and to determine their concentration.

6 Conclusion and perspectives

The target laboratory at IFIN-HH is dedicated to isotopic thin film preparations for various types of nuclear physics experiments with accelerators. The equipment available allows target production by means of thermal evaporation, e-gun, e-beam, sputtering and cold rolling. Additionally, we produce targets by PLD deposition in collaboration with INFLPR. Special care is taken about target quality, which has continuously improved, correlated with our experience and laboratory features. The target characterization in terms of thickness, uniformity, purity, surface morphology and roughness is achieved by means of RBS, SEM/EDX, AFM, and XRD.

With these promising results we have started target production for other international facilities. We are now looking forward to develop a database on thin film characteristics, which would help in refining the aforementioned techniques.

Progress in target characterization is also foreseen along with implementation of the first nuclear forensics laboratory in Romania, which will have modern, high tech equipment, in agreement with ITWG (International Technical Working Group on Nuclear Forensics) recommendations.

To conclude, the manufacturing process combined with characterization techniques lead to innovative target development methods.

This work was supported by the National Institute for Physics and Nuclear Engineering – "Horia Hulubei" (IFIN-HH).

References

1. N. M. Florea, L. Stroe, R. Marginean, D. G. Ghiţă, D. Bucurescu, M. Badea, C. Costache, R. Lica, N. Mărginean, C. Mihai, D. V. Moşu, C. R. Niţă, S. Pascu, T. Sava, J Radioanal Nucl Chem, **305**, 707–711 (2015)
2. A. Stolarz, J Radioanal Nucl Chem, **299**, 913–931 (2014)
3. E.H. Kobisk and H.L. Adair, Nucl. Instrum. Meth. **167**, 153 (1979)
4. J. P. Greene and G. E. Thomas, Nucl. Instrum. Methods Phys. Res., Sect. B **61**, 575-579 (1991)
5. J.M. Heagney and J.S. Heagney, Proc. World Conf. of the INTDS, A-6850-C, 92 (1977)
6. https://www.mantisdeposition.com/nc/systems
7. http://www.durston.co.uk
8. http://www.specac.com
9. Coating services by MRC, Materials Research GmbH, February (1981)
10. I. Burducea, M. Straticiuc, D.G. Ghiţă, D.V. Moşu, C.I. Călinescu, N.C. Podaru, D.J.W. Mous, I. Ursu, N.V. Zamfir, Nucl. Inst. Methods Phys. Res., Sect. B, **359**, 12-19, (2015)
11. M. Mayer, SIMNRA User's Guide, Technical Report IPP 9/133, Garching, Germany (1997)
12. https://www.surrey.ac.uk/ati/ibc/files/IBAPitfalls_NewHandbook.pdf
13. http://www.panalytical.com/XPert3-MRD-XL.htm
14. H.M. Smith and A.F. Tuner, Appl. Opt. **4**, 147, (1965)
15. E. Morintale, C. Constantinescu, M. Dinescu, Physics, AUC, **20**, 43-56 (2010)
16. D. B. Chrisey and G. K. Hubler, Pulsed Laser Deposition of Thin Films, John Wiley & Sons Inc., New York (1994)
17. A.Mitu, M. Dumitru, R. Şuvăilă, A. Oprea, I. Gheorghe, P. Mereuţă, S. Brajnicov, I. Burducea, N. M. Florea, N. Mărginean, T. Glodariu, M. Dinescu, G. Căta – Danil, Vacuum **161**, 162-167 (2019)
18. A. Mitu, A. Oprea, M. Dumitru, N. M. Florea, T. Glodariu, R. Şuvăilă, C. Luculescu, N. Mărginean, M. Dinescu, Gh. Căta-Danil, J Radioanal Nucl Chem, **31**, 725 (2018)
19. T. Rauscher, Phys. Rev. C **73**, 015804 (2006)
20. T. Rauscher, N. Dauphas, I. Dillmann, C. Fröhlich, Zs. Fülöp, Gy. Gyürky, Rep. Prog. Phys, **76**, 066201 (2013)
21. A. Oprea, T. Glodariu, D. Filipescu, I. Gheorghe, A. Mitu, M. Boromiza, D. Bucurescu, C. Costache, I. Cata-Danil, N. Florea, D. G. Ghita, A. Ionescu, N. Mărginean, R. Mărginean, C. Mihai, R. Mihai, A. Negret, C. R. Niţă, A. Olacel, S. Pascu, C. Sotty, R. Şuvăilă, L. Stan, L. Stroe, A. Serban, I. Stiru, S. Toma, A. Turturica, S. Ujeniuc, AEPJ Web of Conferences **146**, 01016, (2017)
22. A. Mitu, M. Dunitru, F. Dumitrache, N. Mărginean, R. Şuvăilă, C. R. Niţă, M. Dinescu, Gh. Căta-Danil, U.P.B. Sci. Bull., Series A, **80**, Iss. 3 (2018)
23. S. Leoni B. Fornal, N. Mărginean, M. Sferrazza, Y. Tsunoda, T. Otsuka, G. Bocchi, F. C. L. Crespi, A. Bracco, S. Aydin, M. Boromiza, D. Bucurescu, N. Cieplicka-Orynczak, C. Costache, S. Călinescu, N. Florea, D. G. Ghiţă, T. Glodariu, A. Ionescu, Ł.W. Iskra, M. Krzysiek, R. Mărginean, C. Mihai, R. E. Mihai, A. Mitu, A. Negreţ, C. R. Niţă, A. Olăcel, A. Oprea, S. Pascu, P. Petkov, C. Petrone, G. Porzio, A. Şerban, C. Sotty, L. Stan, I.Ştiru, L. Stroe, R.Şuvăilă, S. Toma, A. Turturică, S. Ujeniuc, C. A. Ur, Phys. Rev. Lett. **118**, 162502 (2017)
24. T.A. Utigard, M. Wu, G. Plascencia, T. Marin, Chemical Engineering Science, **60**, 2061-2068 (2005)

Preventing damage to floating foils caused by Rayleigh-Taylor instabilities

John O. Stoner, Jr., *Robert B.* Stoner, and *Constance G.* Stoner

ACF-Metals, PMB 231, 2954 N. Campbell Avenue, Tucson AZ, U.S.A. 85719-954

Abstract. An evaporated metal foil target is often produced on a layer of water-soluble parting agent previously applied to a massive substrate. The foil is then floated onto a water surface by immersing the substrate into a water bath. and is picked up later if the foil survives. During the foil's release, a significant fraction of the dissolved parting agent may remain close to the floating foil, as a "heavy" thin layer of solution having higher density than water. This layer of parting agent solution and the lower-density water bath below it form a gravitationally unstable configuration known as a Rayleigh-Taylor instability. If the foil is sufficiently thin, its mass and elastic properties can be ignored, and the motion of the liquids is determined by only the liquids' properties. This system can spontaneously adjust itself toward stability in several ways, one of which involves rotating a cylindrical liquid cell having a horizontal axis, and its cylindrical surface tangent to the surface. This motion moves part of the heavy layer from the top surface downward. The target maker detects this occurrence by the motions of the foil floating on the top of the bath; if the foil is frail, these motions may result in the crumpling, wrinkling, or tearing of the foil. We have observed such behavior with aluminum foils having thickness of 37 nm and diameter of 920 mm on NaCl parting agent, and have successfully implemented methods to prevent such damage.

To use free-standing thin metal foils with thicknesses below one micron, it is often necessary to release (float) them from their substrates and mount them on frames. We found that sufficiently large and thin foils were destroyed by convection during this floatoff. No one had warned us about this. So we had to learn how to prevent it.

We made evaporated metal foils on evaporated water-soluble parting agent previously applied to massive substrates. The foils were aluminum, on sodium chloride parting agent on plastic sheet substrates.

These foils were thinner and larger than average; they were intended to be 37 nm thick (10 $\mu g/cm^2$) and 920 mm in diameter, to investigate the feasibility of foil-neutralizing large-diameter beams of H^- ions. The substrates were Lexan squares approximately 1.2 m x 1.2 m (Fig. 1). Each foil required cleaning one Lexan sheet substrate, evaporating a uniform layer of 300 nm of NaCl parting agent, then evaporating a uniform layer of 37 nm of Al. This process required about three days to manufacture each foil.

After evaporation, each foil was floated onto a water surface by suspending the foil's substrate in a wooden tub lined with polyethylene sheet, having dimensions approximately 1.22 m x 2.5 m x 1.22 m deep, and slowly raising the water level (Fig. 2). If the foil survived, it was then picked up onto a large flat nickel mesh having approximately 10 wires/cm, on a ring support, with 920mm inner diameter. The mesh on its ring waited under water for the foil to be floated; then the water level was lowered by siphoning to lower the foil onto the mesh.

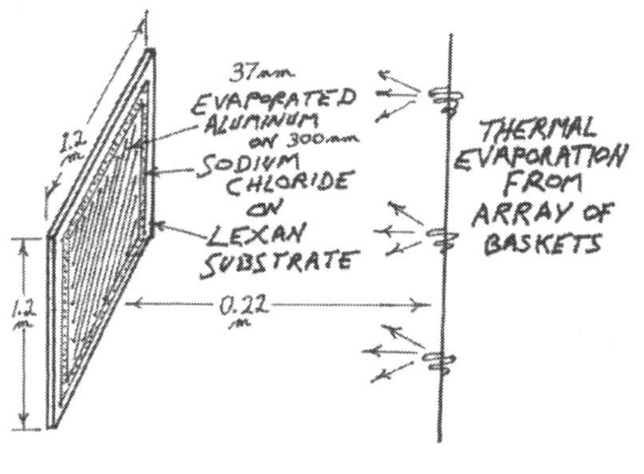

Fig. 1: Simplified sketch of evaporation method for NaCl and Al.

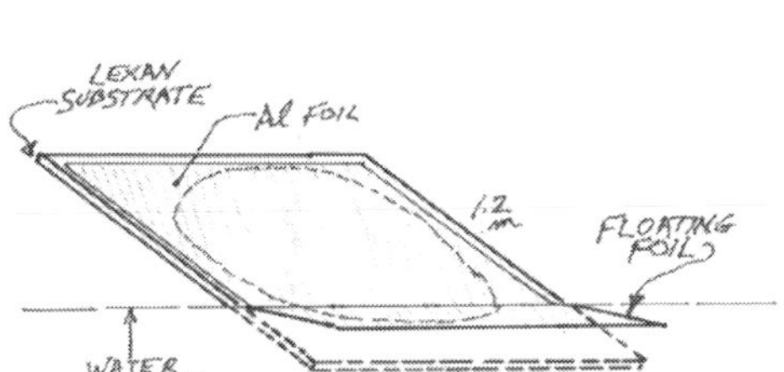

Fig. 2: Floatoff geometry.

 https://doi.org/10.1051/epjconf/202022903002

We needed only one successful foil. However, we had to make eight of them to get one success; the others all failed unexpectedly by collapsing during floatoff. We finally hypothesized that the problem arose from the parting agent. During the foil's release, a significant fraction (perhaps fifty percent) of the dissolved parting agent remained close to the floating foil, as a "heavy" thin layer of solution having higher density than water (Fig. 3).

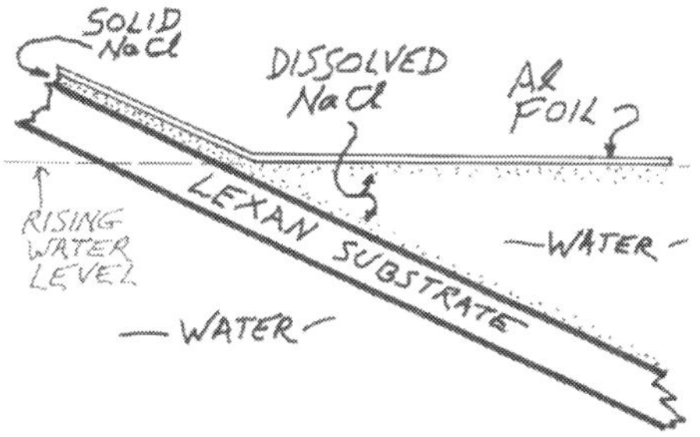

Fig. 3: Detail of parting agent during floating.

The upper layer of parting agent solution on top of the lower-density water bath formed a gravitationally unstable configuration known as a Rayleigh-Taylor instability (Fig. 4) [1].

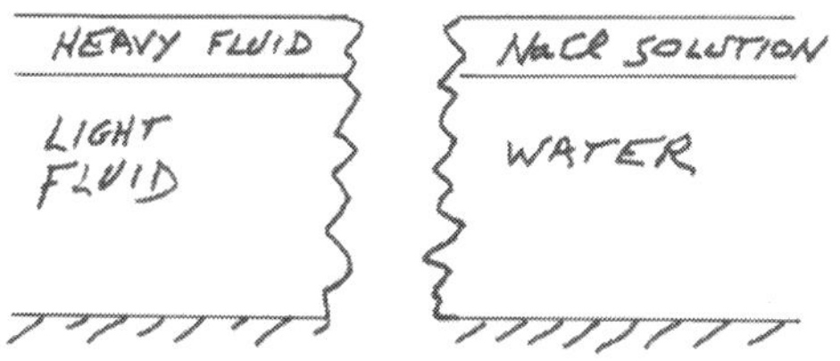

Fig. 4: A Rayleigh-Taylor instability consists of a relatively dense fluid layer above a less-dense fluid layer.

The foil was sufficiently thin that its mass and elastic properties could be ignored, so that the motion of the liquid was controlled primarily by the liquids' properties. Any relatively dense fluid layer remaining slightly above the Lexan substrate was presumed to have little effect. This system could spontaneously adjust itself toward stability in several ways, one of which involved rotating a cylindrical liquid convection cell having a horizontal axis, with its cylindrical surface tangent to the water surface. This motion moved part of the heavy layer from the top surface down to the bottom (Fig. 5).

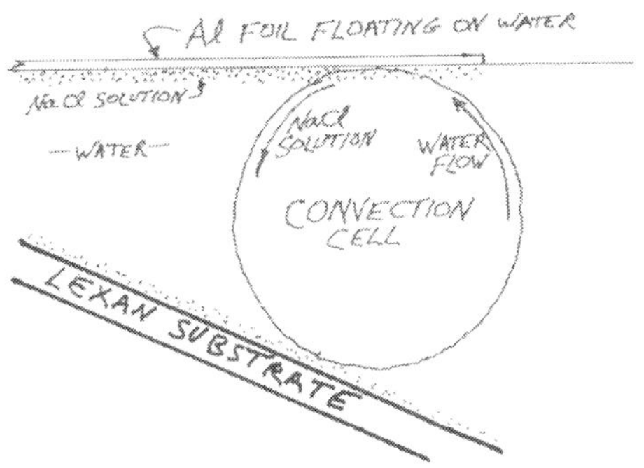

Fig. 5: Convection in a parting-agent-water system.

The target maker detected this occurrence by sudden motions of the foil floating on the top of the bath; the foil was sufficiently frail that these motions resulted in the crumpling, wrinkling, and/or tearing of the foil (Fig. 6).

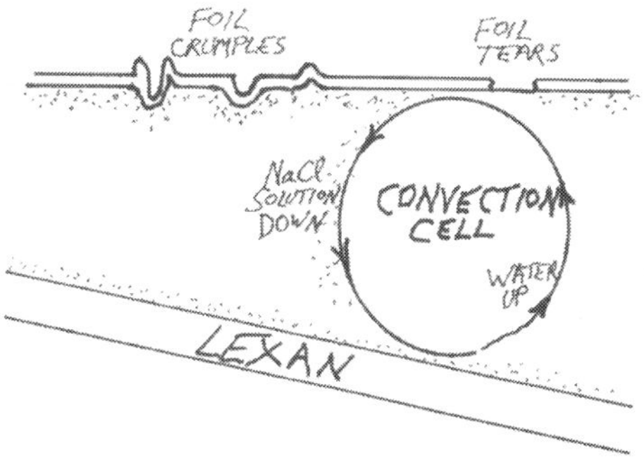

Fig. 6: Damage to floating foil caused by convection.

Fig. 7 shows an example of a foil that demonstrated tearing and collapsing, or crumpling. This is an aluminum foil, about 920 mm diameter, thickness about 37 nm, evaporated onto 300 nm of sodium chloride parting agent. Immediately after it had been completely floated this foil suddenly collapsed in less than 10 s, staying on the water surface. It also tore at several places and broke out a large area, before it was captured on its mesh.

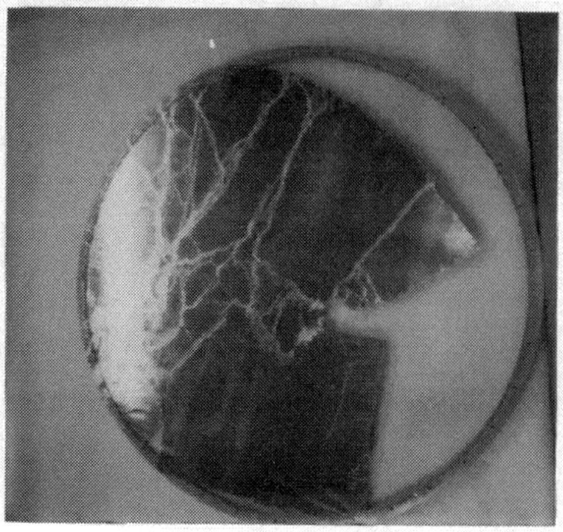

Fig. 7: Foil that wa8.5s torn and collapsed by convection prior to its pickup on a mesh.

We considered several other mechanisms that might cause crumpling or collapsing, and could rule out most of them:

A contaminant film spreading on the water surface might compress the floating foil. But no such film could be found.

Surface tension irregularities might occur due to localized impurities on the Lexan substrate's surface. Pre-washing the Lexan showed no obvious impurities, making this unlikely.

Cooling of the top surface of the float water by evaporation could easily cause convection, simply due to density differences, or also involving temperature-dependent surface tension differences. To minimize the likelihood of these effects, we started each float process by putting half a ton of ice in the bottom of the float tank, to cool the bottom water and make certain that the water in the tank was always warmer toward the top.

The water and the parting agent solution were transparent so we could not directly see convection cells move. We decided that if we could stop them, we did not need to make them visible, so we placed a thin flat "barrier" Lexan sheet a few centimeters below the foil (Fig. 8). As the level of the float water was raised to release the Al foil from its substrate, the barrier sheet was moved toward the substrate. This reduced the coupling of the heavier NaCl solution to the lighter water in the float tank.

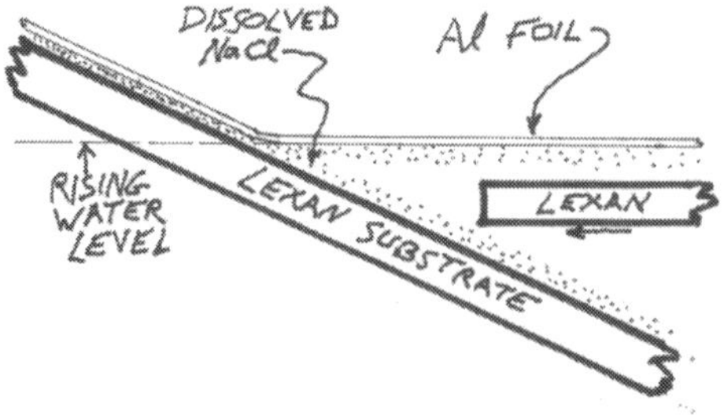

Fig. 8: "Barrier" Lexan sheet placed to prevent large convection cells from reaching the floating foil.

This solved the problem – no more sudden catastrophes occurred during the floating process. We were then able to float off and pick up foil #8, this 920-mm diameter Al foil, without any crumpling or tearing (Fig. 9).

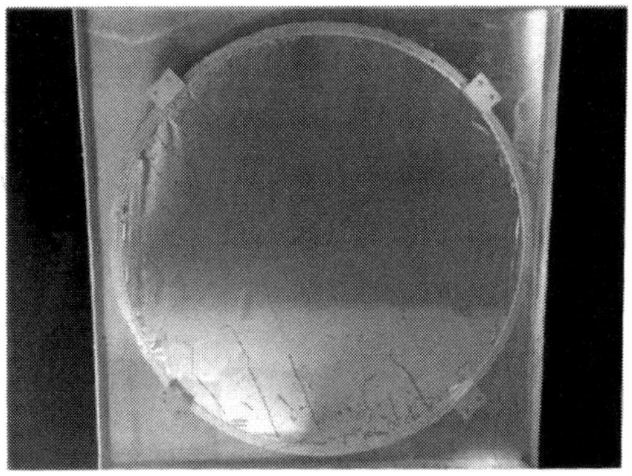

Fig. 9: A foil that was floated and mounted on a mesh without damage by convection.

We concluded that, although convection of the float water represented a serious threat to preparation of large thin foils, the threat could be contained by careful placement of barriers preventing large convection cells from reaching the floating foils.

References

1. D.H.Sharp, Physica *D: Nonlinear Phenomena*, **12**, 1-3, 3-18 (1984)

https://doi.org/10.1051/epjconf/202022903003

The production of deuterated polyethylene targets

N Kheswa[1,*], *TRS* Dinoko[1,2], and *M* Wiedeking[1]

[1]iThemba Laboratory for Accelerator Based Sciences, Somerset West, 7129, South Africa
[2]National Metrology Institute of South Africa, P/B X34, Lynnwood Ridge, Pretoria, 0040, South Africa

Abstract. In this contribution, we describe the method used to produce deuterated polyethylene (C_2D_4) targets. The modification of method based on polyethylene dissolution in hot xylene resulted with stable thin C_2D_4 targets.

1 Introduction

In experimental nuclear physics research transfer reactions (particle pickup or stripping) are a powerful tool to unravel single-particle structures in nuclear systems. Traditionally, experiments utilize transfer reactions in normal kinematics, where light beam particles (such as protons, deuterons, alphas, etc) are accelerated and impinged on heavier target atoms. Unfortunately, the inability to manufacture targets made of radioactive elements with short lifetimes (in a range of few h) limits application of classical set-up i.e. heavy target/light projectile in this type of studies.

However, with many new radioactive ion beam facilities already operational and other being currently under construction, inverse kinematic reactions are becoming increasingly important for nuclear physics experiments. Deuterated polyethylene foils, which are easy to handle are widely used in inverse kinematics measurements. Experimental requirements often seek kinematic energy losses of the beam in the target foil to be minimized, while the targets also have to be able to withstand radiation damage and melting due to the bombardment by incident heavy ions. While for many experiments, a thin target is desirable, others may require much thicker one to increase the signal strength and other deuterated materials, such as deuterated polystyrene. Deuterated polyethylene targets are favourable because they are more resilient to radiation damage and have a higher melting point. Such targets were previously prepared at other laboratories, for example by dissolving polyethylene in warm xylene [1] or using a hot presser [2]. In this work, we report the production of C_2D_4 thin foils applying modified Arnison method [1].

2 The method development

The natural polyethylene powder supplied by Sigma Aldrich was used for method development. The relative density at 25 °C of the polyethylene was 0.92 g/ml with the average molecular weight (Mw) of ~ 4000 and average in molecular weight was (Mn) ~1700. Initially, the polyethylene powder was added to hot xylene and when dissolved it was then transferred directly onto glass slides whilst still hot. In another attempt, the dissolved hot polyethylene was allowed to cool down to room temperature before pouring onto glass slide or dipping the slide in already cooled polymer solution. It was found that both prepared solutions upon cooling separate into two layers with the crystallising polymer at the bottom layer.

In another tests, reactants, the polymer and xylene were heated at the same time. And once the solution reached the boiling point, the hotplate was switched off. The polymer solution was allowed to cool down to room temperature before pouring it onto the glass slides. This solution also formed two separate layers as previously mentioned. When hot solution was poured directly on glass substrate, the white crystalline on substrates was observed upon drying, Figure 1(a). This polymer film could not be released from the substrate even after heating the substrates on the hot plate. The homogenous PE in xylene solution was finally obtained when the solution was stirred continuously during the heating and the cooling; Figure 1(b) presents PE layer/film obtained from homogeneous solution.

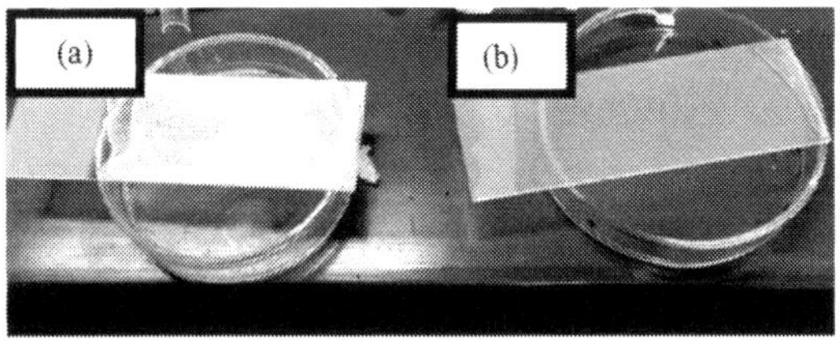

Fig. 1. Glass slides with a natural polyethylene showing the crystallisation of polymer (a) from the solution that formed two separated layers and the slide on (b) is from a homogeneous solution, no. 3

3 Production of deuterated polyethylene targets

The deuterated polyethylene (C_2D_4) material is supplied at relatively high costs compared to natural polyethylene, therefore C_2D_4 targets were reported to be produced in a controlled environment to implement cost saving measures [1, 3, 4, 5]. The method reported by various authors involved dissolving polyethylene in xylene [1, 3, 4, 5]. From these reports, warm, hot or boiling xylene was reported to be used and some authors reported using parting agent on substrates or just polished substrates with no parting agent. In other instances, C_2D_4 targets were reported to be covered with thin film of carbon to enhance their longevity under the beam bombardment [3]. Nevertheless, in all these papers the heat treatment of the deposited polyethylene for the polymerization and evaporation of xylene is reported as the crucial step during the production process. Therefore, we report on the modified method that produced positive results and generated strong targets that were required for nuclear physics experiment to be performed at iThemba LABS. The details of the chain configuration of the deuterated polyethylene (C_2D_4) also supplied by Sigma Aldrich are not available on the supplier's specification sheet, however the isotopic enrichment of the polymer was 98 %.

The C_2D_4 powder of about 1 g was dissolved in 40 ml xylene in a glass beaker. The beaker with the constituents was covered with the watch glass and was placed inside the glass petri dish in order to distribute heat evenly during heating. This was placed on top of the hot plate; see Figure 2(a). The temperature of the hot plate was increased slowly to near the boiling point of xylene (~145 °C) and then kept in this temperature for further 5 minutes. This was then followed by the temperature decrease of 50 °C for every 30 minutes until the hot plate was switched off completely. The solution was let to cool down to room temperature. It should be noted that the polymer solution was continuously stirred during the heating and the cooling down processes. Due to this continuous stirring process, a homogeneous solution was formed in contrast to the two-layered solution which was obtained when the solution was not stirred. Based on the Demaret observation that C_2D_4 polymerizes at 70 °C, at this work the polymer solution was poured onto the slides placed on the hot plate set at 75 °C for polymerisation [5], Figure 2(b). The polymer formation on a glass slide can be observed, surveying changes of the deposit from cloudy solution into transparent solid. The slides with transparent C_2D_4 were scraped on the edges and placed flat in a petri dish filled with boiling water for releasing the polymer. The polymer floated easily to the surface of the water without any parting agent used. The solution can be saved and used for several days with similar results.

4 Discussion and results

Solutions 1 and 2 were obtained during the early developments of the method. Both solutions obtained when PE was dissolved in hot xylene, which formed two layered solution upon cooling, with a polymer at the bottom.

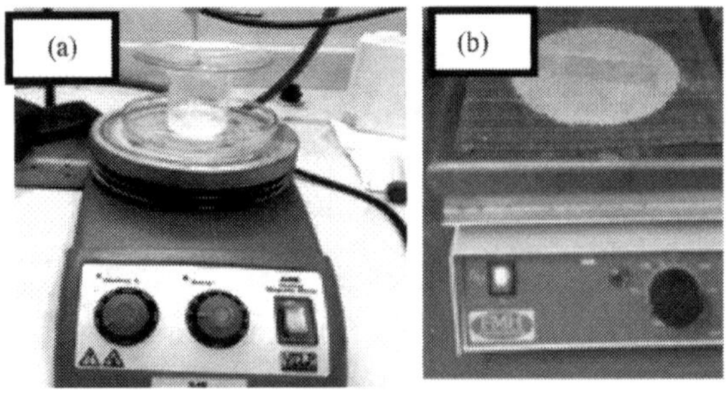

Fig. 2. The images showing (a) the heating processes used during the production of C_2D_4 targets by dissolving the powdered deuterated polyethylene in xylene and (b) the polymerization of C_2D_4 on a heated glass slide

Stirring during the heating and cooling down process of the polymer solutions yielded homogeneous solutions. Homogeneous solutions 3 to 5 presented in Figure 3 were obtained using this procedure. Solutions, 3 and 4 contain PE and 5 is the C_2D_4 solution. The heating of the substrates with deposited polymer for polymerization and use of boiling water for the release of polymer film from the substrate influenced production yield of strong targets which lasted for almost 5 hours of bombardment with a beam. Experiments were targets were used at involved beams of ^{84}Kr and ^{132}Xe projectiles with energies of 338 MeV and 530 MeV respectively.

Fig. 3. Beakers containing PE (1-4) and C_2D_4 dissolved in xylene (5)

5 Conclusion

The method was successfully developed for the production of C_2D_4 targets. One of the crucial aspects of the method was the stirring the solution components (polyethylene and xylene) during heating needed for PE dissolving and during cooling down of the obtained solution what in turn yielded a homogeneous polymer solution. The heating of the glass substrate at the polymerisation of PE and C_2D_4 and using boiling water also improved the step/process of release of polymer film from the glass substrate without the application of any parting agent. Targets with thicknesses ranging from 0.1 to 3 mg/cm^2 were produced with this method. The thickness of targets was determined by weight area measurement. The 1 mg/cm^2 targets reported in this paper, are already used in the nuclear physics experiment conducted at iThemba LABS.

The presented work is funded by the Department of Science and Technology of South Africa through the National Research Foundation.

References

1. G.T.J. Arnison, Nucl. Instr. And Meth. **40**, 359 (1966)
2. M. Kusuhara, Nucl. Instr. And Meth. M, **83**, 328 (1970)
3. G.E. Tripand, B.L. White. Rev. Sci. Instrum. **38**, 435 (1967)
4. H. Fraiquin, INTDS newsletter, 21#**2**, 13 (1993)
5. P. Demaret, INTDS newsletter, 19#**1**, 2 (1992)

Encapsulated Sulfur targets for light ion beam experiments

Sandile Jongile[1,2*], *Ntombizonke* Kheswa[2], *Paul* Papka[1,2], *Olivier* Sorlin[4], *Antoine* Lemasson[4], *Mathis* Wiedeking[2], *Christiaan* Brits[1,2], *Sinegugu* Mthembu[2,5], *Bonginkosi* Kheswa[2,3], *Lucky* Makhathini[1,2], *Fhumulani* Nemulodi[2], *Bongani* Maqabuka[2,6] and *Makuhane* Sithole[2,6]

[1] Stellenbosch University, Private Bag X1, Matieland, Stellenbosch 7602 South Africa
[2] iThemba Laboratory for Accelerator Based Sciences, Old Faure Road Faure, Somerset West, PO Box 722 Somerset West, 7129, South Africa
[3] University of Johannesburg, Auckland Park, 2006, South Africa
[4] Grand Accélérateur National d'Ions Lourds (GANIL), Boulevard Henri Becquerel, 14000 Caen, France
[5] University of Zululand, Private Bag X1001, KwaDlangezwa 3886, South Africa
[6] University of the Western Cape, Robert Sobukwe Rd, Bellville, Cape Town, 7535

Abstract. A new method was developed to produce enriched Sulfur targets with minimum loss of material. This was made possible by inserting Sulfur in-between two 0.5 µm Mylar foils ($C_{10}H_8O_4$). The initial aim was to ensure that the Sulfur targets reduce by no more than 50% of the initial thickness within 24 hours under the equivalent of 10 J/cm^2 of integrated energy deposition by an energetic (E_b > 50 MeV) proton beam. There is no loss of enriched material while making the target, as all the material is deposited within the surface area to be exposed to the beam. During beam irradiation, the targets were frequently swivelled in order to expose each part of the target to the beam and achieve homogeneous irradiation. Targets of 0.4 mg/cm^2 thickness were produced and characterised using ion beam analysis technique with a 3 MeV proton beam.

1 Introduction

Sulfur targets for energetic light ion beam experiments using direct reaction measurements pose certain challenges. Elemental Sulfur targets cannot be produced as self-supporting and this element sublimes rapidly in vacuum when subjected to energy deposition during beam bombardment. When less abundant Sulphur isotopes ($^{33,\,34,\,36}$S) are required financial implications may limit the amount of material that is available. For instance, with only 0.01% natural abundance ^{36}S material is on the higher end of the price list of enriched materials especially if very high purity is required. Previously Sulfur targets were made available in compounds namely Cadmium, Silver, Mercury and Antimony Sulfide [1]. These types of targets were successfully used for neutron-rich nuclei studies in the N=20 region [2-5]. For direct reaction experiments such as (d,p) transfer reactions heavy contaminants are highly undesirable as they lead to many states from the substrate being populated making it difficult to distinguish the states of interest from the contaminants. Furthermore, the cross sections for scattering of the projectile are high which is also detrimental to charged particles detectors. To this respect, a method developed by Hogenbirk *et al.* [6] consisted of sandwiching evaporated Sulfur between evaporated layers and encapsulated between carbon foils. In this paper an alternative method to Sulfur evaporation was investigated to make the best use of enriched Sulfur material using direct powder deposition between Mylar foils. This development was undertaken to perform a nuclear physics measurement using the ^{36}S(p,d) reaction at El_{ab} = 66 MeV with a high resolution magnetic spectrometer at iThemba LABS [7].

2 Experimental Techniques and Equipment

2.1. Production of targets

The targets were produced by inserting Sulfur in between two thin Mylar ($C_{10}H_8O_4$) foils. The interesting part of this method is that while making the target there is no loss of material as it is all deposited within the surface area to be exposed to the beam. Two Mylar foils of 80 µg/cm^2, corresponding to 5.5×10^{18} atoms/cm^2, were "welded" together. This was done by applying heat on the Mylar using a temperature controlled copper cylinder. The temperature was adjusted so that the Mylar foils stick together and encapsulate 1 mg of Sulfur material with a surface area of approximately 1 cm^2 (1.9 ×10^{19} atoms/cm^2). The target frame consisted of two identical rings to support the Mylar foils as displayed in Figure 1 for examples of target before (left) and after

* Corresponding author: sandilej@tlabs.ac.za

(right) irradiation. The rings offer attachments for the swivelling system displayed in Figure 2.

Fig. 1. 1 mg of Natural Sulfur deposited over a disk of 1 cm diameter. Left: target before irradiation, Right: target after 2 continuous days of irradiation with E lab = 3 MeV, I_b = 6 nA proton beam.

Mylar is a good material for encapsulating foils as the contaminants peaks observed are solely from C and O in the Mylar which are well accounted for and therefore do not interfere with our measurement. This was verified during EBS (Elastic Back Scattering) measurements where no discernible structure was observed between the proton energy corresponding to the oxygen elastic peak and the beam energy.

2.2 Elastic Back Scattering characterization

Elastic Back Scattering was performed at the iThemba LABS Material Research Department's (MRD) Tandetron facility. A 3 MeV proton beam was incident on a target made of natural S. A single silicon barrier detector was placed at Θ_{lab} = 135° relative to the beam axis. The natural S target was connected to two independent motors and suspended using springs. The motor and gearbox were mounted within the chamber with connecting rods pulling the target frame with a direction forming a 90° angle with respect to each other. The target is held by one or multiple springs pulling opposite to the motors as shown in Figure 2.

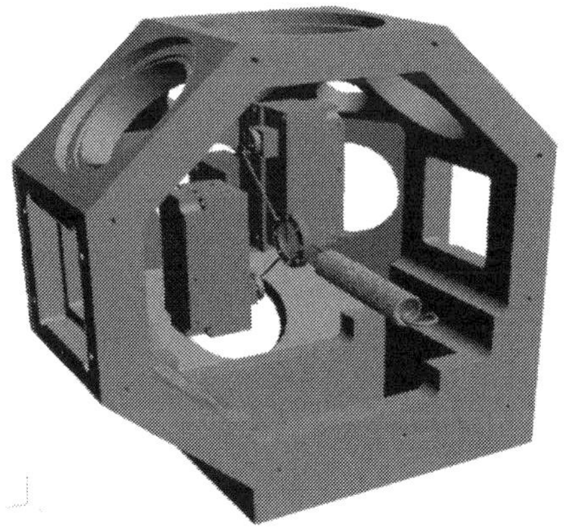

Fig. 2. Two-axis swivelling system to uniformly expose the target to the beam and hence energy deposition. The beam traverses the chamber from left to right. The sides and blanking flanges of the vacuum chambers are deleted from the 3D drawing for viewing purposes.

The target was moved regularly in sequences of 15 minutes in a swivelling motion to expose each part of the target to the beam and achieved homogeneous irradiation while Sulfur is continuously evaporated and sublimated within the space between the Mylar foils. The aim of the test was to evaluate the lifetime of how long the targets last when exposed to a 3 MeV proton beam and with the knowledge that less energy is deposited at a proton energy of 66 MeV allows for an estimating for the target lifetime for 66 MeV protons.

3 Results

The main drawback of this method is the varying target thickness (due to uneven distribution of material in the target), that can be detrimental using low energy beams. This can have implications in the overall energy resolution and can also be detrimental when performing cross section measurements. However, this method allows to employ extremely small amounts of some of the most expensive enriched materials.

Figure 3 displays the EBS spectrum recorded with a fresh target and after the equivalent of 12 hours of irradiation on the area containing natural Sulfur material. A SIMNRA analysis [8] indicates a very high roughness of the encapsulated layer. As the powder is deposited, the thickness varies substantially from areas with no material at all to areas with clusters or large grains. The overall thickness is measured as 450 μg/cm^2 from the fresh target. A relatively large roughness is needed to reproduce the Sulfur tail, and the single C and O peaks otherwise two distinct peaks for C and O would be visible with no roughness. This also produces the low-energy tail on the Sulfur elastic peak.

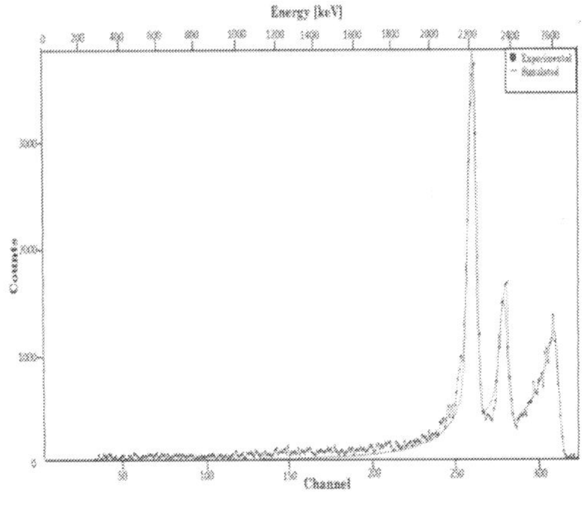

Fig. 3. EBS spectra obtained using the SIMNRA simulations comparing with experimental data.

Assuming a Gaussian distribution of the thickness, the FWHM roughness is 2 mg/cm^2. Considering the final application to this specific target using a 66 MeV proton beam, the resulting degradation of the energy resolution is expected to be δE = 15 keV through 2 mg/cm^2 of Sulfur material which is considered acceptable. It is not clear why half of the areal density is measured using EBS technique as compared to the expected 1 mg/cm^2

from the initial weighted material. It is conceivable that some material is lost during the "welding" process as heat is employed to 300° C for a short period of time. Some of the target material may also not be accounted for because of the large roughness and the possible related errors in the simulations. The targets lasted for approximately 48 hours of continuous irradiation with the 3 MeV proton beam.

Figure 4 shows the scalar spectrum obtained by integrating the Rutherford Back Scattering peak on Sulfur. The beam current was kept constant within +/- 10%. The Sulfur content decreased by 25% over the irradiation period. Considering only the heat deposition, it follows that the energy deposited as per SRIM [9] calculations by 3 MeV protons in 0.4 mg/cm^2 of natural Sulfur is 33.5 keV while that of the 66 MeV protons is 3.2 keV. The time it takes to lose 50 % of the target material is then given as follows:

$$Days = (dE(3MeV)\times 6)/ (dE(66 \; MeV)\times 10) \quad (1)$$

Where:
• dE(3MeV) - energy loss for 1 proton at 3 MeV in 0.4mg/cm^2 of S

• dE(66 MeV) - energy loss for 1 proton at 66 MeV in 0.4mg/cm^2 of S

Therefore assuming that only energy deposition is responsible for loss of material, it is possible to run for 6 days before losing 50 % of the target.

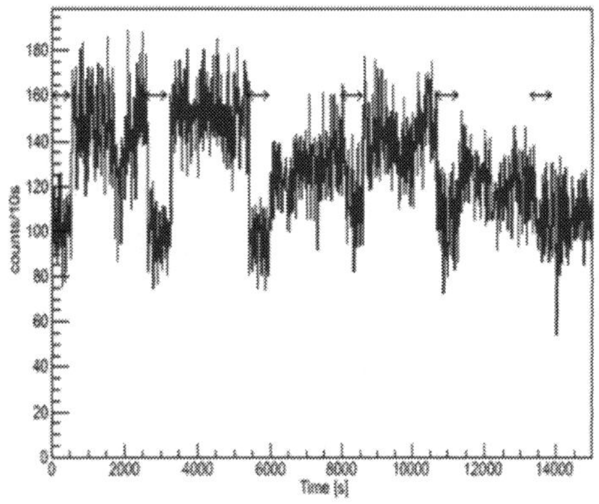

Fig. 4. Arrows indicate when the target is swiveling within region comprising Mylar-Sulfur and Mylar only areas. Target positions are selected when high Sulfur spots are found. The scaler plot is obtained for protons measured with energy E_p between 2.45-2.65 MeV. The accumulated time is 4 hours at I_b = 6 nA, E_b = 3 MeV.

4 Conclusion

The method of encapsulating Sulfur between two Mylar foils is an effective way to produce targets with an extremely small amount of material (0.5-1 mg/cm^2). Good tenure to heat deposition is observed using thin Mylar foils, but the main drawback is the relatively large roughness (FWHM = 2 mg/cm^2) which makes it unsuitable for nuclear reactions with heavy beams or low energy projectiles. Some of the material is not accounted for using EBS analysis. It is theorized to be lost during the welding process or missed because of the large roughness. Further improvements to reduce the roughness are envisaged by applying heat to the Sulfur during the encapsulation, however caution should be taken to avoid loss of Sulfur material.

This work is based on research supported wholly funded by the National Research Foundation of South Africa.

References

1. L.K. Fifield, N.A. Orr, *et al.*, Nucl. Instr. and Methods in Physics Research A**288**, 360-363 (1990)
2. C.E. Thorne, J.W. Olness, E.K. Warburton and S. Raman, Phys. Rev. C**30**, 1442 (1984)
3. P.V. Drumm, L.K. Fifield, R.A. Bark, M.A.C. Hotchkis, C.L. Woods and P. Maier-Komor, Nucl. Phys. A**441**, 95 (1985)
4. L.K. Fifield, C.L. Woods, R.A. Bark and M.A.C. Hotchkis, Nucl. Phys. **A440**, 531 (1985)
5. L.K. Fifield, C.L. Woods, W.N. Catford, R.A. Bark, P.V. Drumm and K.T. Keoghan, Nucl. Phys. **A453**, 497 (1986)
6. A. Hogenbirk, H.P. Blok, M.G.E. Brand, A.G.M. van Hees, J.F.A. van Hienen and F.A. Jansen, Nuclear Physics **A516**, 205 (1990)
7. R. Neveling *et al.*, Nucl. Instr. and Methods **A654**, 29 (2011)
8. M. Mayer, *SIMNRA User's Guide*, Report IPP **9**/113, Max-Planck-Institut für Plasmaphysik, Garching, Germany (1997)
9. J F. Ziegler *et al.*, Nucl. Instr. and Methods **A268**, 11-12 (2010)

Preparation and characterization of ^{10}B targets at JRC-Geel

David Vanleeuw*, *Jan* Heyse, *Goedele* Sibbens, and *Mariavittoria* Zampella

European Commission, Joint Research Centre, Directorate G, Retieseweg 111, Geel B- 2440, Belgium

Abstract. Measurements of neutron-induced cross sections to generate nuclear data are a core activity of the JRC-Directorate G Standards for Nuclear Safety, Security and Safeguards unit in Geel. Thin ^{10}B layers are of great importance in this activity as they are used to measure the absolute neutron flux in the beam using the standard $^{10}B(n,\alpha)^{7}Li$ reaction cross-section as a reference. After a period of reduced activity and in line with a renewed interest for nuclear data, the demand for high quality ^{10}B targets increased. In this paper we describe the design and features of a new e-beam evaporator specifically customized for the preparation of boron targets as replacement of the old dysfunctional equipment. Several ^{10}B targets of varying thicknesses were prepared and characterized as part of the factory acceptance tests and implementation in the JRC-Geel target preparation laboratory. Differential substitution weighing was applied for mass determination and in order to calibrate the thickness monitor. Comparative time-of-flight measurements relative to ^{10}B and ^{235}U standard targets were conducted at the GELINA neutron time-of-flight facility at the JRC-Geel site as second methodology for the determination of ^{10}B areal density. The morphology of the layers was assessed by means of Scanning Electron Microscopy (SEM). The determination of impurities was realized by means of Energy Dispersive X-ray (EDX). Finally, two boron targets were prepared in the frame of the measurement of the neutron induced fission cross-section of ^{230}Th at the n_TOF neutron time-of-flight facility at CERN.

1 Introduction

The target preparation laboratory of the JRC in Geel was established in 1961. The first boron targets were prepared in 1967 by high frequency induction [1] and later on by e-beam evaporation for the preparation of boron reference standards for the neutron lifetime experiment [2]. By means of a Leybold Univex 450 e-beam evaporator, a set of ^{10}B reference targets was prepared and characterised by isotope dilution mass spectrometry and by irradiation in a thermal neutron beam. Despite the need of high quality and well characterised boron targets, the target preparation activities were stopped during the nineties due to a reorganisation of the JRC [3]. After the nuclear renaissance, the renewed interest for accurate nuclear data resulted in a new demand for high quality targets in order to provide the neutron facilities of the JRC in Geel with the necessary targets [4,5]. More specifically, high quality ^{10}B targets were requested for the absolute neutron flux monitoring of the beam by means of the $^{10}B(n,\alpha)^{7}Li$ reaction cross-section data as standard reference.

In view of this renewed interest, the target preparation activities were revived [6] and in a first stage the old Leighbold Univex 450 e-beam evaporator was put back into service in 2009 after years of inactivity. Several boron targets were prepared with this equipment but poor signal after exposure to the neutron beam questioned the quality of the layers. The mass of the layers was therefore monitored at several time intervals with a microbalance and by means of comparative weighing. This indicated a clear loss of mass in time which demonstrated the instability of the deposited layer, most likely due to the poor adsorption of the boron atoms as the original operational conditions could no longer be reached. This problem continued to persist despite the variation of parameters such as substrate type, deposition time and boron material. As neither the maximal power of the e-beam nor the vacuum in the chamber could be improved it was concluded that new equipment was required in order to comply with the demand for high quality boron targets. Efforts were made to acquire new highly enriched boron material but due to a lack of producers the boron stock of the target preparation laboratories was used.

In this paper we discuss the technical feasibility, acquisition and implementation of new deposition equipment for thin boron films. The technical features of the equipment are explained in relation with the specific needs for the preparation of boron targets and in view of emerging techniques in the field of thin layer deposition. The preparation of several boron targets of varying thicknesses is explained from factory acceptance tests (FAT) and start-up on up to the preparation of two large boron targets requested under a collaboration agreement with CERN in the frame of a neutron induced fission cross-section measurement of ^{230}Th at the n_TOF facility. Two techniques are applied for the determination of the boron and ^{10}B areal density of the layer and the calibration of the thickness monitor. The results of the Scanning Electron Microscope (SEM) and

* Corresponding author: david.vanleeuw@ec.europa.eu

Energy Dispersive X-ray (EDX) analysis for assessment of the layer quality and elemental composition are presented and discussed as well as further needs for characterisation.

2 E-beam evaporation

2.1 Deposition technique

Depending on whether the process is primarily chemical or physical, thin film deposition techniques fall into two broad categories, chemical vapor deposition or physical vapor deposition (PVD). Based on the acquired experience, the characteristics of boron and the requirements of the deposited layer in terms of purity and geometry, physical vapor deposition by thermal evaporation stood out as the most suitable technique. PVD processes are atomistic deposition processes in which material with high melting point and low vapor pressure is vaporized from a solid source in the form of atoms or molecules and transported in the form of a vapor through a vacuum or low pressure gaseous environment to the substrate, where it condenses. Thermal evaporation is a PVD process whereby the source material is thermally heated by means of resistance heating of a boat or by high energy electron beam (e-beam) heating of the source material itself and is used as a deposition technique for thin layers with a thickness in the range of a few nanometers to thousands of nanometers. The high melting point (2349 K) and heat of vaporization (508 $kJ \cdot mol^{-1}$) of crystalline boron requires thermal evaporation by e-beam heating. When the high-speed electrons hit the boron in the water cooled crucible, sufficient energy is generated to melt and convert the boron material into vapor and then deposit it onto the substrate. The source to substrate distance can roughly be determined according to the required homogeneity (cosines law for a non-spherical surface) and yield (mean free path of molecules). In the case of higher flux processes such as electron-beam evaporation, collisions between evaporant atoms can lead to a focusing of the vapor plume [7]. Due to the low vapor pressure and hence low evaporation rate there are however no problems expected with the uniformity of the boron vapor.

2.2 Deposition equipment

The technical specifications for the e-beam evaporator were based on the experience with existing setups and on an evaluation of potentially interesting new features in order to gain more control over the deposition process. The distance to the single substrate stage was set at 40 cm (like in the previous setup), but it was extended with the possibility to move the substrate closer to the evaporation source for improved yield when thicker layers are requested (and decrease the deposition time as boron has a very low vapor pressure), or to increase the distance to guarantee homogeneity when larger samples are requested. The substrate stage can rotate with a speed up to 40 rotations per minute and can hold a substrate with a maximum diameter of 12 cm. A stainless steel mask holding the substrate is mounted underneath the rotating stage and needs to be customized according to the requested spot size and the dimensions of the substrate. Frequently, very thin aluminised foils are requested as backings for the ^{10}B deposit in order to minimise energy loss of the emitted reaction particles in the backing. The fragility of these backings makes the cleaning prior to deposition difficult. In order to remove physically adsorbed hydrocarbon contamination from the surface of these backings which might interfere in the deposition process (adsorption of the boron atoms), it was decided to place a glow discharge module in front of the substrate stage. By means of ionized argon gas, the substrate can be cleaned prior to the deposition as the energetic plasma removes the weakly adsorbed contaminants [8,9]. Furthermore, the possibility to heat the substrate during and after deposition was added to the technical specifications in order to have control on the homologous temperature, which is the relation between the evaporation and sample temperature and is known to influence the morphology and quality of the layer according to the Structure-Zone Model of thin films [10].

After publication of the technical specifications and according to the European rules for public procurement, the contract for the development and construction of a new customized e-beam deposition system was awarded to Moorfield Nanotechnology®. The contract consisted of two phases where the construction phase was preceded by the conceptual phase (Fig. 1). In this phase, requiring approval prior to the construction phase, extensive collaboration based on the specific needs and practical experience with e-beam evaporation took place between the JRC Geel and Moorfield Nanotechnology®. The FAT consisted of the successful preparation of a boron target.

Fig. 1. 3D-design of the evaporation chamber: quartz heater (1), substrate holder (2), quartz crystal (3), glow discharge (4), shutter (5), evaporation source with e-beam (6), vertical movement substrate (7).

3 Preparation of boron targets

The final quality of the target depends on the quality of the starting material. This means that the material should be as pure as possible, highly enriched in the case of isotopes and preferably single element. For the implementation of the new evaporator we used crystalline boron in the form of lumps with a chemical purity better than 98.0% as declared by the certificate of the provider. No specific enrichment was mentioned and the natural abundance of 19.9% of ^{10}B was assumed with a relative uncertainty of 3% [11]. Several boron targets were prepared as part of the acceptance tests, implementation, optimization and calibration of the equipment, characterization of the areal density and finally for a neutron time of flight experiment.

3.1 Targets for factory acceptance tests

Four test targets were prepared at the premises of Moorfield Nanotechnology® as part of the FAT and in order to establish the deposition parameters. The starting material was loaded in three of the six pockets protected with a tungsten liner in the cupper crucible. The 0.5 mm thick aluminium substrates with a diameter of 50 mm were degreased with methanol and weighed by means of a Mettler-Toledo XP56 microbalance prior to deposition in view of applying differential weighing for the determination of the deposited mass. The height of the rotating substrate stage was set in the lowest position and a mask with a diameter of 30 mm was used to mount the substrate and define the spot size. All evaporations were conducted at a pressure better than 10^{-5} Pa which was a significant improvement compared to the old Leighbold evaporator (10^{-2} Pa). All features could be tested and approved (shutter, rotation of the pockets, rotation of the sample, vertical movement of the sample, heating of the sample). Each deposition run included prior in-situ cleaning with the argon plasma (10 minutes at gas flow of 30 $cm^3 \cdot min^{-1}$).

The optimal settings of the evaporation source were defined during the first evaporation runs by gradually increasing the power of the e-beam until a significant signal was registered by the thickness monitor (several $Å \cdot s^{-1}$) and while applying high frequency sweeping. This occurred at the maximal current of 200 mA and resulted in a grey deposit, different from the expected appearance of a boron deposit as observed on reference boron targets prepared at JRC-Geel in the past. After further investigation, a small cavity could be observed in the tungsten liner indicating the evaporation of tungsten instead of only boron due to the high power of the e-beam. During the next two deposition runs, more attention was paid to the heat distribution of the e-beam by applying lower current (40 mA) and gently moving the e-beam instead of high frequency sweeping over a large area, resulting in a low deposition rate more in line with the low vapor pressure of boron. The two targets prepared under these conditions (FAT 3 and FAT 4) visually resembled the expected appearance of a boron deposit and were sent to JRC-Geel for mass determination and SEM/EDX analysis.

3.2 Targets for implementation and calibration

After confirmation of the quality of the last two FAT targets, the e-beam evaporator was installed in the JRC-Geel target preparation laboratory. In view of further testing and implementation of the equipment as well as calibration of the thickness monitor, five more test deposits (Test 1 to 5) were prepared according to the previously defined deposition parameters on the same type of substrates and with the same spot diameter. In order to prepare deposits with increasing mass, five deposition runs ranging from 30 to 120 minutes were conducted at a pressure better than 5.10^{-5} Pa and an e-beam current between 30 and 40 mA. During the five deposition runs, the deposition rate ($Å \cdot s^{-1}$) and final thickness (Å) according to the thickness monitor were monitored. The settings of the thickness monitor were adapted according to the properties of boron as recommended by the Inficon manual (acoustic impedance of 22.699 $Pa \cdot s \cdot m^{-1}$ and density of 2.370 $kg \cdot m^{-3}$) and the unknown tooling factor was kept at 1 (see further). The final thickness of the layers was also determined by differential weighing of the masses of the deposited layers and by comparative neutron time-of-flight measurements at the GELINA facility.

3.3 Targets for cross section experiment

Finally, two targets were prepared in view of neutron induced-fission cross-section measurements at the neutron time-of-flight facility in CERN. The substrates consisted of 0.025 mm foils of aluminised Mylar with a diameter of 120 mm mounted on a aluminium ring with a thickness of 1 mm and inner diameter of 100 mm. These foils are too fragile to clean mechanically and only in-situ plasma cleaning was applied. The size of the substrate and the very low requested areal density of 8 $\mu g \cdot cm^{-2}$ on an 80 mm spot size also excluded accurate differential weighing. The first target was therefore prepared according to the calibrated thickness monitor and measured by comparative time-of-flight measurement. The second target was prepared according to these results.

4 Characterization

4.1 Calibration thickness monitor

In order to deposit layers of a predefined thickness as defined by the target requestor, a reliable real time measurement during evaporation is required. The most common equipment for this purpose and also applied for the e-beam evaporator, is a quartz microbalance connected to a thickness monitor. A mass variation per unit area can be measured by the change in frequency of the water cooled quartz crystal resonator. As mass is deposited on the surface of the crystal, the thickness increases and consequently the frequency of oscillation decreases. This relationship was quantitatively established by Sauerbrey in 1959 [12] and further improved so that by introducing parameters inherent to the deposited material, such as density (the quartz crystal only measures the areal mass density) and acoustic impedance, the frequency change can be quantified and correlated precisely to the film thickness.

The thickness monitor measures however only the amount of material deposited on its sensor, not how much is deposited on the substrate. To avoid shadowing effects, the sensor cannot be placed in view of the substrate and may not even be at the same distance from it. Therefore, the rate at which the material is deposited on the sensor may not equal the rate at which it is deposited on the sample. The ratio of the two rates is called the "tooling factor (TF)" F_m and can be determined as follows:

$$F_m = F_i \cdot (T_m/T_i) \quad (1)$$

where F_i is the initial TF, T_i is the film thickness indicated by the instrument, and T_m is the actual, independently measured thickness of the deposited film. If no tooling factor has been pre-set or used before, F_i equals 1. Accurate weighing of the deposited layer with well-defined area is used as measurement method for the thickness of non-radioactive layers in the JRC-Geel target preparation laboratory [13] and was also applied for the boron targets. A second independent method was applied as confirmation.

4.2 Differential weighing

The methodology for thickness measurements of non-radioactive targets in the JRC-Geel laboratories consists of differential weighing of the target before and after deposition by means of substitution weighing according to the International Organization of Legal Metrology (OIML R111-1 : C.4 1-2) and defining the surface area of the deposit (spot) by applying a mask with well-known dimensions. Substitution weighing is a highly accurate and precise mass determination through comparison with mass standards to guarantee traceability to the SI units. Temperature, air pressure and humidity are measured for air buoyancy correction.

A Mettler-Toledo XP56 microbalance and a set of E2 calibration weights were used for all the weighing. Each target was weighed three times before and after deposition according to the SUUS method of substitution measurement where the Standard (S) and Unknown (U) are weighed in sequence after each other. The final masses were calculated after correction for air buoyancy and the combined standard uncertainties with a coverage factor k = 1 estimated in accordance with the Guide to the Expression of Uncertainty in Measurement [14]. The major uncertainty components result from the uncertainties on the mass standards, the balance parameters, the weighing and the diameter of the mask.

Obviously, the mass of the deposited layer takes all deposited elements into account, including other isotopes as well as the co-deposited impurities. If the impurities in the starting material are also deposited (which might happen due to the high evaporation temperature of boron, apart from the outgassing products before deposition which are retained by the shutter), the mass obtained by the differential weighing needs to be corrected for the purity of the boron and the enrichment in order to calculate the areal density of the ^{10}B.

Finally, differential weighing was also applied to verify the stability in time of the prepared boron targets.

4.3 Neutron time-of-flight measurement

A second independent characterization method for the amount of ^{10}B in the deposited film was applied. As previously mentioned, differential weighing takes all elements of the layer into account, so a method only measuring the amount of ^{10}B is more relevant. The second method consisted of mounting the prepared boron targets back-to-back with a fixed ^{10}B target on the cathode of a double, gas-filled, Frisch-gridded ionization chamber [15], and exposing them to one of the neutron beams of the GELINA neutron time-of-flight facility [16]. The fixed ^{10}B target is used as a neutron fluence monitor, in order to link consecutive measurements to each other.

As the $^{10}B(n,\alpha)^7Li$ reaction occurs, an α- and a ^{7}Li-particle are emitted in opposite directions. One of both charged particles will be emitted in the direction of the target backing, in which it will be stopped. The other particle will be emitted in the gas volume of the chamber, where it will ionize the gas due to interactions with the gas molecules. By applying an external voltage to the anode and cathode, the electrons and positive ions resulting from the ionization will start moving in opposite directions and induce a charge on both electrodes. When the electrodes are connected to an external read-out system, the induced charge will produce an electronic signal which can be registered. The amplitude of the signal will be proportional to the energy of the ionizing particle (i.e. α or ^{7}Li).

The reaction cross section, which is a measure for the reaction probability, is well-known for the $^{10}B(n,\alpha)^7Li$ reaction and considered a standard in the neutron energy range between thermal neutron energy and 1 MeV [17]. As such, the number of reactions registered by either side of the ionization chamber, allows determining the neutron fluence as a function of neutron energy if the areal density of ^{10}B nuclei in the boron target is well-known; or, alternatively, it allows determining the number of ^{10}B nuclei in a ^{10}B target if the neutron fluence as a function of neutron energy is well-known.

Two ^{10}B reference samples with different areal density were used for calibrating the ionization chamber. These samples were characterised by neutron activation in the BR2 research reactor of the SCK-CEN with a well-known neutron fluence in the frame of the accurate measurement of the neutron lifetime [2]. By consecutively putting both reference samples in the ionization chamber, back-to-back with the fixed ^{10}B monitor sample, it was possible to deduce and verify a correlation between the neutron fluence at the position of the reference sample and the response of the detector on the side with the fixed ^{10}B sample.

An additional verification was done by placing a well-known ^{235}U sample in the same position as the reference samples and by using the standard reaction cross section for the $^{235}U(n,f)$ reaction at thermal neutron energy [17].

https://doi.org/10.1051/epjconf/202022903005

The samples produced with the new e-beam evaporator were consecutively mounted in the position of the reference samples. Their areal density for ^{10}B nuclei was determined, using the known neutron fluence deduced from the response of the fixed ^{10}B sample side of the chamber and the standard cross section for the $^{10}B(n,\alpha)^{7}Li$ reaction.

4.4 SEM-EDX

The morphology of the layer of several boron targets was investigated by means of a JSM-7800F Schottky Field Emission Scanning Electron Microscope (SEM) in high vacuum mode. Images were taken in low depth focus at 10X magnification for the general appearance of the sample and in SEM mode at magnifications of 250X and 1000X on several positions. For all the image acquisitions the accelerating voltage was 20 kV and the lower secondary electron detector was used.

An elemental analysis was performed by Energy Dispersive X-ray spectroscopy (EDX) to confirm the presence of boron and check the presence of impurities and more specifically the absence of tungsten from the liner of the crucible. This was done in the center of the sample at a working distance of 10 mm and at different penetration depths by varying the accelerating voltage from 1 to 30 kV. The acquisition time was 60 s and the spectra were normalized. The results are qualitative and the reported weight percentage values are indicative only.

5 Results and discussion

After determination of the appropriate settings of the e-beam (power and sweeping), nine boron targets were prepared (Table 1): two targets for the FAT (FAT3 and FAT4), five targets for the calibration and implementation in the target preparation laboratories (test 1 to 5) and finally two targets in view of a neutron time-of-flight experiment at CERN (TP2017-10-01 and 02).

Table 1 Overview of the boron deposits prepared with the new e-beam evaporator

Target	e-beam current (mA)	Deposited mass (g)	Areal Density, weighing (µg/cm2)	Areal Density B10, weighing and natural abundance (µg/cm2)	Areal Density B10, TOF (µg/cm2)	SEM	EDX
FAT 1 and 2	200	-	-	-	-	-	-
FAT 3	30 -40	0.000142(14)	20.2(8)	4.0(2)		Some dust particles, layer follows Al morphology	B, Al, O, Fe traces, no W
FAT 4	30-40	0.000260(13)	36.8(24)	7.3(5)		Some dust particles, layer follows Al morphology	B, Al, O, Fe traces, no W
Test 5	30-40	0.000039(18)	5.6(25)	1.1(5)			
Test 4	30-40	0.000079(17)	11.1(25)	2.2(5)			
Test 3	30-40	0.000154(18)	21.8(26	4.3(5)			
Test 2	30-40	0.000289(18)	40.8(30)	8.1(6)	7.0		
Test 1	30-40	0.000527(19)	74.5(40)	14.8(9)	12.7	Some dust particles, layer follows Al morphology	B, Al, O, Fe traces, no W
TP2017-10-01	30-35	no weighing	8.0 according monitor	2.8 calculated according monitor	2.9		
TP2017-10-02	30-35	no weighing	22.2 according monitor	8.0 calculated according monitor			

Visually, all deposits show a good layer quality and different colors are observed at different layer thicknesses due to a varying reflection of visible light (Fig. 2). Despite some dust particles, the layer is smooth and follows the roughness profile of the unpolished aluminum substrates as can be seen on the SEM images (Fig. 3). This proofs the good adhesion of the boron layer which is also confirmed by the layer stability, verified by regular weighing. The layers deposited with the new e-beam evaporator are thus clearly of much better quality than those prepared with the old Leybold evaporator.

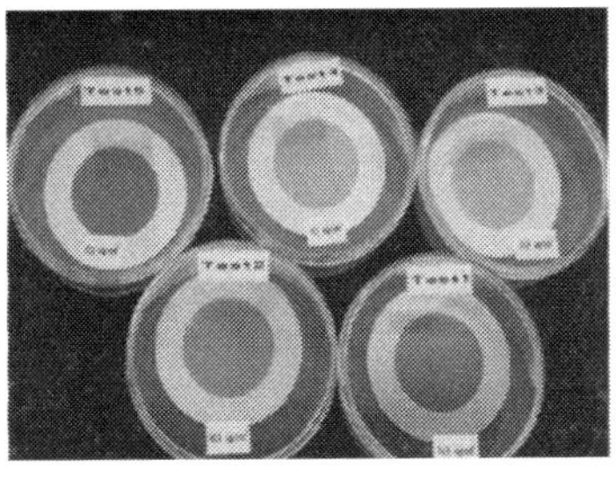

Fig. 2. Pictures five boron test targets prepared at JRC-Geel

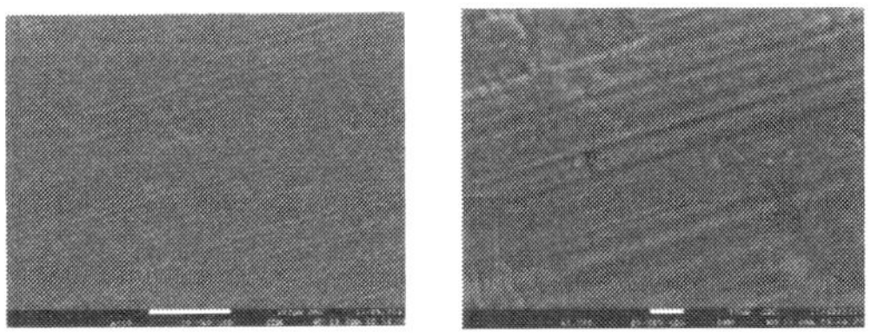

Fig. 3. SEM image of FAT 3 at magnification 250X (left) and 1000X (right) with same centre point

The EDX analysis at different penetration depths (1 kV, 5 kV and 30 kV) confirms the presence of boron and the absence of tungsten from the liner placed in the crucible. The thickness of the boron layer is smaller than theaverage penetration depth at 30 kV as demonstrated by the increase of the boron mass fraction at decreasing acceleration voltages. Aluminum and oxygen were also detected and are most probably present in the form of aluminum oxide in the top layer of the substrate. Metallic aluminum is very reactive with atmospheric oxygen and a 4 nm aluminum oxide layer forms on any exposed aluminum surface [18]. Traces of iron (0.1 wt%) were detected at 30 kV with 10 min acquisition time. The EDX results indicate that the deposited layer is not affected by impurities coming from the starting material; this may depend on the high purity of the starting material and/or on the absence of co-evaporation.

Fig. 4. EDX spectrum of Test 1 taken at the centre of the sample at 5 kV

Table 2 Calculation of the Tooling Factor

B Target	Measured thickness monitor (Å)	Measured total mass layer (g)	Calculated areal density (μg/cm2)	Calculated thickness (nm)	Calculated Tooling Factor
test 5	125	0.000039(18)	5.6(25)	24(11)	188(85)%
test 4	250	0.000079(17)	11.1(25)	47(10)	188(43)%
test 3	500	0.000154(18)	21.8(26)	92(10)	184(24)%
test 2	760	0.000289(18)	40.8(30)	172(11)	227(20)%
test 1	1500	0.000527(19)	74.5(40)	314(11)	210(15)%

The thickness of the boron layers deposited on the aluminum substrates as determined by differential weighing for a spot size with a diameter of 30 mm as defined by the mask corresponds to an areal density varying from 5.6 to 74.5 μg·cm^{-2}. For the five test targets, these results were compared with the readouts of the thickness monitor (Fig. 5) and show a significant linear correlation ($p<0.05$, $r = 0.997$), meaning that within an uncertainty range the thickness monitor can be used to deposit a predefined thickness. Besides for the direct preparation of a specific layer thickness, this is of particular importance for very thin layers on large substrates unsuitable for thickness determination by weighing.

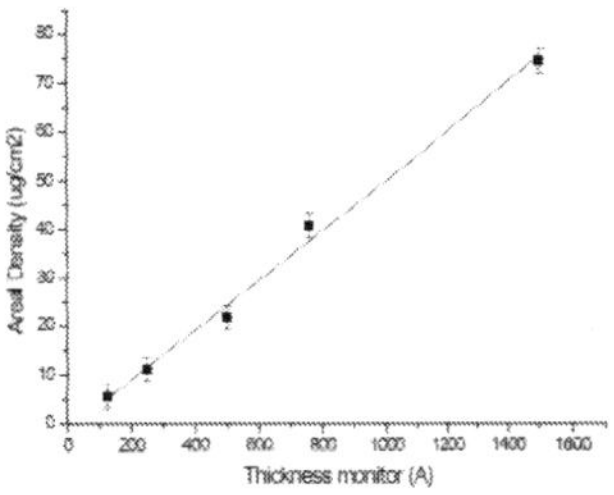

Fig. 5. Graphic representation of the readout of the thickness monitor and the areal density according to weighing

After conversion of the areal density to the thickness of the boron layer by means of the density of metallic boron, the TF for each deposit was calculated according to equation 1 (Table 2). We notice an important fluctuation of the TF which in this case can be explained by the large uncertainty on the weighing results of the thin boron layers used to calculate the layer thickness. As by experience the TF appears to be more stable when evaporating powder by resistance heating and when taking a closer look at the boron material during and after evaporation, this fluctuation might also find its origin in shadowing effects or the non-uniform evaporation. Due to the properties of metallic boron, the evaporation takes place more locally where the e-beam hits and heats up the crystalline boron lump. This creates moving craters in the material and continuously modifies the geometry and hence TF, which should be a fixed value with fixed geometry of the sample size and height, evaporation source and crystal.

As no impurities are detected by EDX (above the detection limits and except for 0.1 wt% Fe) and no enriched boron was used, the areal density of ^{10}B can directly be calculated with the natural abundance of ^{10}B (Table 1). A second method was applied for direct determination of this value by comparing the response of a double Frisch-gridded ionization chamber for the test targets 1 and 2 with the response of two reference boron targets with well-known areal density of ^{10}B (originating from the neutron lifetime project [2]) and a ^{235}U sample with a well-known areal density. Both targets were mounted consecutively in the ionization chamber, back-to-back with a fixed ^{10}B sample used as neutron fluence monitor, and put in one of the neuron beams of the GELINA neutron time-of-flight facility. Preliminary results of these measurements, point to an amount of ^{10}B which is 15% lower than expected. An in-depth analysis of the results obtained with the time-of-flight technique, including an uncertainty analysis, is ongoing. With an uncertainty on the weighing of 3 to 6%, the preliminary results might point to an overestimation of the amount of ^{10}B based on the weighing and natural enrichment. This might be due to the presence of undetected impurities in the boron layer (below the detection limit of EDX) or more probably due to a lower enrichment of the starting material. In order to clarify the reason of this discrepancy, also in view of the use of other batches of enriched boron, it is therefore necessary to assess the purity and the isotopic composition of the different boron starting materials stored for a long time in the target preparation laboratory. This will be done at JRC-Karlsruhe by combining different techniques based on mass spectrometry. The total content of boron in the starting materials will be measured by isotope dilution mass spectrometry (IDMS) with an inductively coupled plasma mass spectrometry (ICP-MS) instrument. IDMS is considered the most precise method for quantitative determinations of the concentration of multi-isotopic elements and ICP-MS is the most used technique for boron determinations among present-day technologies [19]. Due to its low detection limits (at ppb level) ICP-MS will also be used to check the eventual presence of impurities in the boron starting materials by multi-element analysis. The isotopic composition of the boron starting materials will be measured by thermal ionization mass spectrometry (TIMS) which is a technique that can provide a high degree of accuracy and precision of the measurement results [20].

In order to prepare the large ^{10}B deposits with a thickness of 8 $\mu g{\cdot}cm^{-2}$ as requested by CERN, the real time measurement of the thickness monitor according to the previously determined tooling factor of 210% for test 1 was applied for target TP2017-10-01. Due to the large uncertainty on the TF and as the areal density could not be determined by differential weighing, the target was measured relative to the boron standards, resulting in a much lower thickness than expected (Table 1). Closer examination of the positioning of the much larger substrate stage in the evaporation chamber showed a shadowing effect of the quartz crystal resulting in a larger TF than previously calculated. Based on the last results, a new correction factor was applied for the preparation of the second target TP2017-10-02. The latter could not be characterized due to the planning of neutron induced fission cross-section measurement of ^{230}Th.

6 Conclusion

A new e-beam evaporation system customized according to the specific needs for the deposition of thin boron layers and based on the latest available thin layer technology was acquisitioned following the public procurement rules and successfully implemented by conducting several evaporation runs. The prepared boron targets visually resemble the reference boron targets and further assessment by SEM-EDX analysis confirms a smooth and well adsorbed layer without defects or impurities originating from the starting material or evaporation liner. Accurate differential weighing was applied to determine the total areal density and monitor the stability of the layers. Five test deposits of different thicknesses were prepared to calibrate the thickness monitor. The areal density of ^{10}B was determined by comparing these targets with ^{10}B and ^{235}U reference deposits in a double Frisch-gridded ionization chamber and exposing them to one of the neutron beams of the GELINA facility. The results are in line with the expected natural enrichment but the deviation between both characterization methods requires the need of further and more accurate characterization of the starting material. Finally, the acquired experience in terms of preparation and characterization was applied to provide two thin boron targets for CERN in view of a neutron time-of-flight experiment.

After a decade of technical incapability, the target preparation laboratory is once again able to prepare high quality boron targets which are of great importance for the determination of the neutron flux in the frame of nuclear data research.

References

1. J. Van Audenhoven, J. Joyeux, Nucl Inst and Methods **57**, 157 (1967)
2. J. Pauwels, R. Eyckens, A. Lamberty, J. Van Gestel and H. Tagziria, R.D. Scott, J. Byrne, P.G. Dawber, D.M Gilliam, Nucl. Instrum. Methods Phys. Res. A **303**, 133-140 (1991)
3. J. Van Audenhove, J. Pauwels, J. Jaklovsky (Eds.), *Preparation of Nuclear Targets for Particle Accelerators*, 79-87 (1981)
4. https://ec.europa.eu/jrc/en/research-topic/neutron-data Accessed on 02/08/2018
5. A. Plompen, JRC-IRMM, AIP Conf. Proc. **769**, 1417 (2005)
6. A. Stolarz, R. Eykens, A. Moens, Y. Aregbe, Nucl. Instrum. Methods Phys. Res. A **613**, 351–356 (2010)
7. A. Powell, P. Minson, G. Trapaga, U. Pal, Metall. Trans. A **32**, 8, 1959-1966 (2001)
8. Staff of Kaufman & Robinson Inc, Technical Note KRI-06
9. A. Belkind, S. Gershman, *Vacuum Technology & Coating* (2008)
10. J.A. Thornton, Proc. SPIE **0821**, Modeling of Optical Thin Films, 95-103 (1988)
11. S. Szegedi, M. Váradi, C. Buczkó, M. Várnagy, T. Sztaricskai, J. Radioanal Nucl. Chem. **146** (3), 177 (1990)
12. G.Z. Sauerbrey, Phys, **155**, 206–222 (1959)
13. D. Sapundjiev, A. Dean, U. Jacobsson, K. Luyckx, A. Moens, G. Sibbens, R. Eykens, Y. Aregbe, Nucl. Instrum. Methods Phys. Res. A **686**, 75–81 (2012)
14. https://www.bipm.org/en/publications/guides/gum.html
15. C. Budtz-Jørgensen, H.-H. Knitter, Ch. Straede, F.-J. Hambsch, R. Vogt, Nucl. Instrum. Methods Phys. Res. A **258**, 209-220 (1987)
16. W. Mondelaers, P. Schillebeeckx, Notiziario Neutroni e Luce di Sincrotrone **11** (2), 19-25 (2006)
17. A.D. Carlson, V. G. Pronyaev, R. Capote, G. M. Hale, Z.-P. Chen, I. Duran, F.-J. Hambsch, S. Kunieda, W. Mannhart, B. Marcinkevicius, R. O. Nelson, D. Neudecker, G. Noguere, M. Paris, S. P.Simakov, P. Schillebeeckx, D. L. Smith, X. Tao, A. Trkov, A. Wallner, W. Wang, Nuclear Data Sheets **148**, (143-1882018)
18. T. Campbell, R. Kalia, A.Nakano, P. Vashishta, S. Ogata, S. Rodgers, Phys. Rev. Lett. **82** (24), 4866 (1999)
19. R.N. Sah, P.H. Brown, Microchem. J., **56**, 285-304 (1997)
20. R.N Sah, P.H Brown, Biological Trace Element Research **66**, 39-53 (1998)

Uranium targets for heavy-ion accelerators

Bettina Lommel*, *Elif* Celik Ayik, *Annett* Hübner, *Birgit* Kindler, *Jutta* Steiner, and *Vera* Yakusheva

GSI Helmholtzzentrum für Schwerionenforschung, Targetlabor, Planckstrasse 1,64291 Darmstadt, Germany

Abstract. Uranium targets are very important for accelerator-based research of nuclear properties. Depending on the reaction to be studied and on the conditions during the experiments different restrictions on the target material have to be met; as for example, durability, melting temperature, reactivity or a possible contribution of the additional compounds present to the reaction. Therefore, we are developing processes to produce uranium targets in the elemental form as well as in different compounds.
Here we report on the production and application of targets from metallic uranium, UF_4 and UO_2.

1 Introduction

Uranium is a commonly used target for accelerator-based research on nuclear properties. While uranium and uranium tetrafluoride (UF_4) have a melting point significantly above 1000 K, uranium oxide (UO_2) and uranium carbide (UC) have a significantly higher melting point above 2000 K. For high-intensity beams, targets of higher melting compounds are better suited. Metallic uranium oxidizes easily in (humid) air. As most of the commercially available metallic uranium has already some oxygen component, additional oxidation cannot not be hindered, even when kept under inert gas. Therefore, only a target of a high-melting uranium compound, which is stable in air, guarantees for long-lasting stability.

2 Starting material

Uranium and all its compounds are highly toxic, radioactive, accumulate in inner organs and endanger the environment. From uranium, only radioactive isotopes are known. In the target laboratory at GSI, we apply only ^{238}U. The radioactive laboratory of the target laboratory has an allowance for uranium, which is depleted from natural uranium and has a depletion of below 0.4 % ^{235}U. For our own safety, we work with material depleted lower than 0.2 % ^{235}U, typically. Beside the pure metal, the carbide, fluoride and oxide are an option because of their better stability in air. In the following, we describe the different starting materials and their availability on the market.

2.1. Metallic uranium

Metallic uranium can be obtained as a sputter target or as evaporation material from Framatome [1], Manufacturing Sciences Corporation [2], or SC Chepetsky Mechanical Plant [3]. Physical vapour deposition from a boat or from a crucible is difficult, as uranium forms numerous compounds, and therefore crucibles are destroyed easily. Physical vapour deposition with the electron-beam gun has a very low yield. Although this method was applied in the past [4], methods with less wasteful consumption of radioactive materials are preferable. Metallic uranium targets on backings can be obtained by DC-magnetron sputtering [5].

2.2 Uranium tetrafluoride (UF_4)

Uranium tetrafluoride (UF_4) can be deposited with physical vapour deposition from the tantalum boat or from the tantalum crucible [5]. The material is stable in dry air. The material is no longer commercially available in Europe, but we still have remaining stock bought in the past from Merck Company [6].

2.3 Uranium oxide (UO_2)

Uranium oxide (UO_2) can be deposited with physical vapour deposition from the electron-beam gun. Physical vapour deposition with the electron-beam gun has a very low yield. Although this method was applied in the past [7], methods with less wasteful consumption of radioactive materials are preferable. We succeeded in obtaining uranium oxide as a hot pressed 1"-sputter target.

2.4 Uranium carbide (UC)

Uranium carbide (UC) is an interesting target material, as well. Uranium carbide has a high melting point of $T_m = 2500$ K. It is available in different compounds, UC would be the material of our choice, since the other carbides are difficult to be obtained in a stoichiometric compound and some decompose during heating. A sputter target out of UC is commercially not available.

* Corresponding author: B.Lommel@gsi.de

3 Physical vapour deposition

In the target laboratory at GSI electron-beam gun deposition, thermal deposition, resistance heating, DC-magnetron sputtering and RF-magnetron sputtering are the methods available for target fabrication. Deposition with an electron-beam gun has a yield of about 5 % for homogenous layers. This is not an option for the deposition of radioactive material. Thermal deposition is an option for uranium tetrafluoride with its melting point of T_m = 1300 K and crucibles out of tantalum or platinum are available. Uranium oxide and uranium carbide can only be processed with an electron-beam gun or a sputtering set-up. Insulating materials are sputtered with RF-magnetron sputtering; conductive materials are sputtered with DC-magnetron sputtering. For materials with a known poor electrical conductivity, the sputter ability with a DC-magnetron has to be tested experimentally.

3.1 Thermal deposition of UF_4

Uranium tetrafluoride (UF_4) can be deposited with physical vapour deposition from the tantalum boat or from the tantalum crucible [5]. We apply a tantalum crucible to deposit up to 500 µg/cm² uranium fluoride on carbon backing up to a diameter of 30 mm, respectively. We obtain uranium fluoride layers with a homogeneity of ± 5 %. The targets are stable in dry air. Depending on the application, the targets are covered with a carbon layer to hinder the sputtering in the beam [5].

3.2 DC-magnetron sputtering of metallic U

Metallic uranium can be deposited by DC-magnetron sputtering from a water-cooled 1“-copper magnetron [5]. We deposit metallic uranium on Be, C, Al, Ti, Nb and Ta backings up to a thickness of 2000 µg/cm² and up to a diameter of 30 mm. We obtain uranium layers with a homogeneity of ± 5 %. The targets oxidize easily under air, even quicker under humid air. Depending on the oxygen content of the starting material the oxidation in an argon glove box cannot be hindered, naturally it slows down but after weeks of storage the uranium is affected as well. Depending on the application, the targets are covered with a carbon layer to hinder the sputtering in the beam. The carbon covering layer does not hinder the oxidation.

3.3 DC-magnetron sputtering of (UO_2)

Uranium oxide can be deposited by DC-magnetron sputtering from a water-cooled 1“-copper magnetron. We deposit UO_2 on carbon backings as well as on titanium backings with a thickness of up to 2000 µg/cm² and up to a diameter of 30 mm. We obtain uranium layers with a homogeneity of ± 5 %. The targets are stable in dry air.

The uranium oxide sputter target is a hot sintered ceramic material with poor thermal conductivity. For the sputtering of uranium oxide a rather high power of about 150 W is needed. Therefore, the process is divided into steps of 2 minutes sputtering time to avoid overheating and to give the set-up time for cooling down after each step.

To obtain a thickness of about 2500 µg/cm² in total 400 minutes of sputtering are needed. A denser sputtering target, which is produced by hot powder metallurgy or another compound, can be an alternative.

3.4 Physical vapour deposition (PVD) of UC

We are planning to obtain uranium carbide as a sputter target. We assume that uranium carbide can be sputtered with a DC-magnetron system.

4 Application of the obtained targets

The uranium oxide targets will be applied in experiments of nuclear structure at GSI in Germany, at ELI in Romania and at SARAF in Israel.

References

1. www.framatome.com
2. mfgsci.com/
3. www.chmz.net
4. H. Folger, W. Hartmann, J. Klemm and W. Thalheimer, Nucl. Instr. and Meth. A **282** 73 (1989)
5. B. Kindler, D. Ackermann, W. Hartmann, F. P. Heßberger, S. Hofmann, A. Hübner, B. Lommel, R. Mann and J. Steiner, Nucl. Instr. and Meth. A **590** 126 (2008)
6. www.merckgroup.com
7. J. G. Povelites, Report AERE-R **5097** (1965)

The Argonne Target Library

Gavin Gordon[1], *Matthew* Gott[1], and *John P.* Greene[1,*]

[1]Physics Division, Argonne National Laboratory, 9700 South Cass Avenue, Lemont, IL, 60439, United States

Abstract. As part of the proposal to DOE-NP for the Center for Accelerator Target Science (CATS) initiative, one of the objectives was to develop an inventory of existing targets that will serve as a pool available to the community. Targets collections have been recovered from Yale University due to the closing of their Tandem Accelerator Facility. In addition, accumulated targets from target preparation in the Physics Division over several decades have also been assembled with the intent of providing them to whomever would have a use for them. Space has now become available to compile, catalogue and house these collections. Thus, the Argonne Target Library has been established and its progress and outlook will be discussed in detail.

1 Introduction and motivation

1.1 Center for Accelerator Target Science (CATS)

Over the past decade, the need for targets for the low-energy nuclear physics program has been expressed in many venues. With the closure of several accelerator laboratories, the demand for targets has only increased on existing target-making facilities. The Physics Division at Argonne National Laboratory (ANL) maintains a target development laboratory in direct support of ongoing low-energy nuclear physics research undertaken at the Argonne Tandem Linac Accelerator System (ATLAS) facility. The mission of the target laboratory is also to support research by the low-energy research group of the Physics Division. Under certain conditions, the target laboratory has also supplied targets to other scientists, including ATLAS users, for experiments carried out at other locations, both within the US and abroad, but only a fraction of these outside requests can be addressed. It was thus proposed to create a National Center for Accelerator Target Science (CATS) based on the existing target development laboratory at ANL. The objectives of the center are as follows:

1. Serve the low-energy community by producing targets whenever possible by either manufacturing them or by directing requests to other facilities best able to perform the task;

2. Train individual investigators and students in target making in order to provide a workforce capable to address present and future needs;

3. Carry out R&D activities dedicated to novel production techniques and optimization of existing ones;

4. Develop an inventory of existing targets that will serve as a pool available to the community.

Over the past several decades, a very large number of targets have been produced and could possibly be re-used if they were easy to locate. Sletten, in 1974, described a target catalogue system and library used at the Niels Bohr Institute [1]. This CATS proposal would explore the feasibility of generating an inventory of targets from several institutions and made available for reuse.

1.2 The closing of the Yale Tandem

Upon the closure of the Wright Nuclear Structure Laboratory (WNSL) at Yale University [2] there arose the need to preserve many of the target collections already housed there. Targets from both Brookhaven National Laboratory (BNL) as well as Los Alamos National Laboratory had been stored and cataloged at WNSL, occupying several cabinets. These, along with the numerous targets prepared over the years, were recovered and shipped to ANL as part of the initial target library collections.

1.3 The Physic Division Target collection

While the targets collected from Yale have constituted a good start to the Argonne Target Library, there exists an accumulation of accelerator targets prepared by the Physics Division Target Laboratory. Over a time span of thirty plus years these targets have been in use, reuse and just stored for eventual disposal. Now, with the advent of the target library, space is available for sorting and cataloging this large collection of targets for reuse by others. The size of this collection has resulted in an ongoing effort which is still underway.

* Corresponding author: greene@anl.gov

2 The Argonne Target Library

Work has begun assembling this inventory of existing targets that will serve as a pool available to the low-energy nuclear physics community on a first-come first-serve basis. Target collections stored at Yale University consisting of targets recovered from BNL and LANL, targets prepared at Yale and added to those already existing at ANL will form the initial holdings of the Argonne Target Library.

2.1 Space establish

First needed would be a space large enough for the shelves and cabinets needed to hold the collections with some thought as to future space for additional target collections as they might arise. With generous assistance for the Argonne Physics Division a large room, formerly part of the 6-MeV Van de Graff accelerator facility, was made available for this purpose. The area required decommissioning and extensive clean-up. With the equipment and storage removed from the room, the floor was resealed and the windows and blinds cleaned. An office desk and laboratory bench were not discarded and instead kept for eventual library use.

2.2 Furnishings found

Throughout the Physics Division Building can be found spare furnishings and shelve units available for re-use. Several storage cabinets were also located and moved into the newly recovered space. A schematic of the Target Library layout is shown in Fig. 1. Additional furnishings consisting of a computer desk and office chairs, etc. were quickly assembled. A network line was needed and desktop computer needed for target listings was installed.

2.3 The Physics Division Target Archives

Ancillary to the Target Library and therefore co-located within are the Physics Division Target Archives. The totality of target preparation notes dating back to 1963 constitute these archives. Over the years these collections of target preparation notes have been stored in various locations within the Physics Division, including the basement. Every target job and accompanying writings are held in a series of binders containing the target request and notes detailing the target preparation techniques employed, including information on isotope and assays [3]. After 1978 these target request pages have been digitized and are now contained in a searchable database for instant look-up. The target job information can then be easily found in these collections of binders by Job Number. It is natural for this collection to therefore find a home in the Target Library.

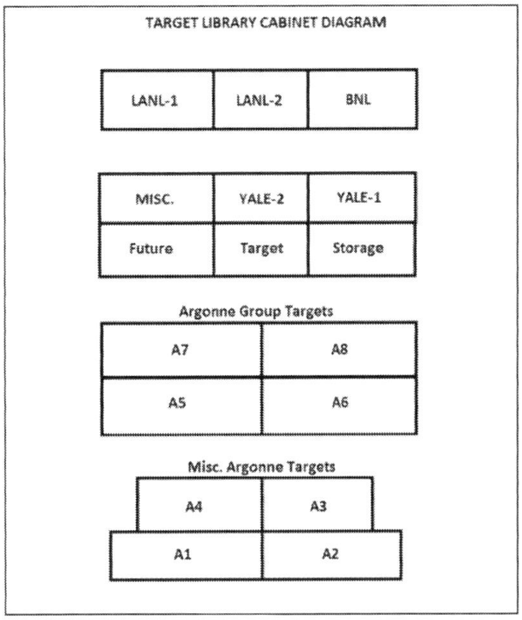

Fig. 1. Schematic layout of the Argonne Target Library

3 The collections

We describe here now the many collections of targets which will constitute the newly established Argonne Target Library.

3.1 The Wright Nuclear Structure Laboratory

As mentioned previously, the Wright Nuclear Structure Laboratory at Yale University had inherited collections of targets from both LANL and BNL when their respective low-energy nuclear physics programs were discontinued. With the closing of the WSNL these individually packaged targets, extensively catalogued, listed in .pdf files (see for example Fig. 2) and adequately stored in a series a of cabinets were carefully packed and shipped to ANL. It was a simple procedure to unpack and reshelf these two collections into the cabinets installed in the target library.

3.2 The Physics Division Target Laboratory

Encompassing a span of over thirty years, many target preparations have been performed within the Physics Division Target Laboratory which resulted in many hundreds of targets made during practice depositions as well as actual targets whereby many are produced to make efficient use of precious separated isotope. This results in many unused targets available for re-use via the target library. They have been stored over the years in the target lab and at various locations within the ATLAS facility.

Stand A

Number	Element	Mass number	Thickness	Backing
1	LiF		200	30 C
2	C		850	-
3	C		20	-
4	C		20	-
5	C	12	~50	-
6	C		10	-
7	C	12	40-50	-
8	N	15	100	10 C
9	Mg	26	~200	-
10	Mg	26	10	20 C + 1 Bi
11	Mg	26	10	20 C + 1 Bi
12	Mg	26	10	20 C + 1 Bi
13	Al		75	Polystyrene, overcoat 10 Au
14	Al	Pure	400	Crème Cote
15	Al	Pure	400	Crème Cote
16	Al		815	-
17	Al		10	-
18	Al		1	-
19	Al	Spot	100	20 C
20	Al		20	overcoat 10 Au
21	Al		75	overcoat 10 Au
22	Al		450	-
23	Al		450	2 Au
24	Al	Pure	400	Crème Cote
25	Al		160	-
26	Al	Spot	81	20 C
27	Al		1	-
28	Si	28	10	20 C
29	Si	28	10	20 C
30	SiO_2		150	-
31	Ca	48	98	20 C
32	Sc		3.85	-
33	Ti	Nat	500	5 mil Ta
34	Ti	Nat	500	5 mil Ta
35	Ti	46	1.02	F=4cm^2
36	Ti	48	1.04	F=3cm^2
37	Ti	50	1	F=3cm^2
38	Mn	Nat	108	Polystyrene
39	Fe_2O_3	Nat	36	20 C
40	Co	59	23	20 C
41	Co		62	20 C
42	Co	59	28	20 C
43	Co	59	28	20 C
44	Co		23	20 C
45	Ni	60	50	20 C
46	Ni		0.00002"	-
47	Ni	60	50	5-10 C

Fig. 2. Example page from the list of BNL targets

3.3 Various miscellaneous collections

In addition to the targets described above, there are many additional targets which have been brought or shipped to Argonne for experiments at ATLAS. Plus there have been many targets produced for Physics Division employees conducting experiments at other facilities throughout the world. These targets are characterized by their originating institution as well as the unique target frames employed. These smaller assemblies of targets were simply entered into the computer as individual lists using Microsoft Word.

4 Current progress and outlook

Much progress has already been made but there is still some unknowns to be determined and work finished up. While the LANL and BNL targets were already nicely catalogued the various other target collections (Yale U., ANL, etc.) and their databases will need to be formulated and built. In the end a working distribution system will need to be established with some simple rules applied. Widespread advertisement and target availability will be accomplished by this publication as well as word-of-mouth through the INTDS and access from the newly established CATS Website.

4.1 Reconstituting the Yale targets

In addition to the carefully listed targets from LANL and BNL recovered from WNSL, there were many targets prepared at Yale over the years and when the lab was closed down, these targets were carefully boxed up and shipped back to ANL. The database chosen for target cataloguing was Microsoft Access which is what is currently employed for the Physics Division Target Archives. Also unexpectedly arrived from Yale were collections of target preparation materials organized by element.

4.2 Cataloguing the ATLAS collections

The various collections of targets accumulated throughout ATLAS and the Physics Division constitute the largest amount of targets to be considered. Here we find many boxes of individual targets as well as groups of targets by experiment, apparatus and investigator. We found it prudent to keep these groups intact and have chosen Microsoft Excel as the simplest way to list these targets. The individual targets were sorted and entered into the computer using Microsoft Access for alphabetical listing.

4.3 Target library operations and CATS website

Once all the various target catalogues are completed, target library operations can commence. So as to prevent individual monopolizing of library requests we are proposing a set of modest rules to be followed:

1. Individual target requests will be limited to one specified target and up to two targets shipped (if possible ± for a spare).

2. Priority will be assigned to domestic requests and of those, being DOE (& NSF) investigators will be addressed first.

3. Wherever possible a user supplied FedEx account will be used for shipping.

4. When necessary, if a vacuum target transport container [4] is needed (such as those marketed by MicroFoils, Inc.) ± It should be supplied in advance by the requestor.

5. Additional considered rules which may present themselves in the future could be approved as necessary.

These library rules and target collection listings will be made available in the near future via the Physics Division CATS Website.

5 Conclusion

In conclusion, the Argonne Target Library is now coming online. Once the catalogued collections are being compiled and completed, they should become rapidly available through a link from the CATS website. We anticipate a slow and steady demand for these targets once established. The mechanism for target check-out should quickly be realized once underway. Hopefully, this vast wealth of unused targets may again become available to the low-energy community for use in nuclear physics experiments.

This material is based upon work supported by the U.S. Department of Energy, Office of Science, Office of Nuclear Physics, under Contract No. DE-AC02-06CH11357. This research used resources of ANL's ATLAS facility, which is a DOE Office of Science User Facility. The authors are grateful to the site planning and area clean-up provided by Kimberly Miller, furniture procurement from Building Manager Keith Trychta and janitorial support from David Streeter.

References

1. G. Sletten, INTDS Conference Proceedings 47 (1974)
2. http://www.yalescientific.org/2015/03/rebirth-of-wright
3. G.E. Thomas, J.P. Greene, Nucl. Instr. Meth. Phys. Res. A **362**, 201 (1995)
4. J.H. Bjerregaard, P. Knudsen, G. Sletten, Book, Ed. : J. Jaklovsky, Plenum Press, NY, 207 (1981)

Method development for producing thin ^{14}C foils

Matthew Gott[1,*], *John* Greene[1], *Igor* Pavlovsky[2], and *Richard* Fink[2]

[1]Physics Division, Argonne National Laboratory, Lemont, Illinois, United States
[2]Applied Nanotech Inc., Austin, Texas, United States

Abstract. Thin, isotopic ^{14}C foils are of great interest to the nuclear physics community as neutron-rich targets. Historically, these foils have been extremely difficult to prepare and an effort is underway to make them readily available. The stock material of ^{14}C available at Argonne contains a number of oxide impurities (SiO_2, MgO, and Al_2O_3), which affect the composition and stability of the fabricated foil. A simple, robust method was developed (using ^{nat}C as a surrogate) to purify the ^{14}C material while minimizing loss and potential spread of the material. Thin foils were fabricated using the purified carbon, the unpurified carbon/oxide mix, and purchased high-purity carbon powder. SEM and EDS of the resulting foils was performed and the efficacy of this purification method was demonstrated.

1 Introduction

There is heavy interest in having low-Z, neutron-rich targets such as ^{14}C, particularly for studying nuclear structure of neutron-rich radionuclides and for astrophysical research experiments. Argonne National Laboratory has a stock of ^{14}C material identified for making nuclear physics targets. This material is legacy material from Los Alamos National Laboratory's pion program and was reported as containing 75% ^{14}C, 16% ^{12}C, and 9% impurities (magnesium oxide, aluminum oxide, and silicon oxide). Preliminary targets made utilizing a mixture of natural carbon with these impurities were found to have some issues with stability and degradation during production. To provide beam-stable, thin ^{14}C target, a purification method was devised to remove the bulk of these impurities prior to preparing the foil. This publication presents some interesting early results in this process.

2 Methods and materials

2.1 Materials

Carbon powder, magnesium oxide, aluminum oxide, and silicon oxide powders were all purchased from Sigma-Aldrich (St. Louis, Missouri). Aluminum foil was purchased from Goodfellow (Coraopolis, Pennsylvania). Sodium hydroxide (50% w/w) and isopropanol were purchased from Sigma-Aldrich and used as provided without further purification. A special binder mixture was prepared for making the carbon foils; a 5% w/w dispersion of ethyl cellulose in 80:20 toluene: ethanol (cP 300; Sigma Aldrich) was further diluted with isopropanol to the desired 2.5% w/w concentration. All water was purified on-site using a Millipore system set to greater than 18 MΩ·cm.

2.2 Removal of impurities

With the ^{14}C stock material containing oxide impurities, a simple, glove-box-friendly method was developed to remove the bulk of these impurities. A homogenized test mixture of bulk, natural carbon with magnesium oxide, aluminum oxide, and silicon oxide added as impurities was prepared and an aliquot was transferred to a 50 mL centrifuge tube. NaOH (50% w/w) was added and gently stirred to incorporate. The tube was shaken several times over the course of a day to ensure interaction of the material with the NaOH and set aside to settle overnight. Gravitational settling was selected over centrifugation due to space limitations for equipment in the glovebox. As the material settled, a white oxide layer settled on top of the bulk carbon; this oxide layer was pipetted away for one sample and left intact for an additional sample. Once the material had settled, the liquid was pipetted off and fresh deionized water was added to rinse away the NaOH and dissolved impurities. This was repeated for a total of three water washes. The material was then pipetted into the pre-weighed polypropylene jars using a small amount of isopropanol for the transfer. This material was allowed to completely dry and the mass of the material was then determined prior to preparation of foils.

2.3 Printing the carbon on an aluminum foil

The foils were printed following a graphene printing method that was recently described in the literature [1]. In short, the dried carbon samples were combined with binder and isopropanol to produce the "carbon paint" for preparing targets. All samples were 1: 1: 0.5 by weight

* Corresponding author: mgott@anl.gov

of carbon: binder: isopropanol. The natural carbon sample used 0.5 g carbon, the purified carbon with oxide removed (PWR) sample used 0.49 g carbon, purified carbon without oxide removed (PWOR) sample used 0.62 g carbon, and the impure carbon sample used a 0.50 g of bulk carbon mixed with the various oxide impurities. Each sample was mixed 3 x 15 seconds in a SpeedMixer at 3000 rpm with 5 minutes of cool down between each run.

Once the paints were prepared, an aluminum foil was set in place on the draw-down machine (Figure 1). 0.5-0.6 mL of the carbon paint was drawn up in a syringe and pipetted across the aluminum foil in front of the Mayer rod. The printer bar was pulled across the surface of the foil to spread a uniform layer of the paint across the surface of the foil. The foils were allowed to air dry for 10 – 15 minutes prior to handling them. Then, each foil was then cut to fit inside of 100 mm diameter petri dishes. The lids were placed, so the remaining organic solvents could escape and the foils were all placed in a 100° C oven for two hours.

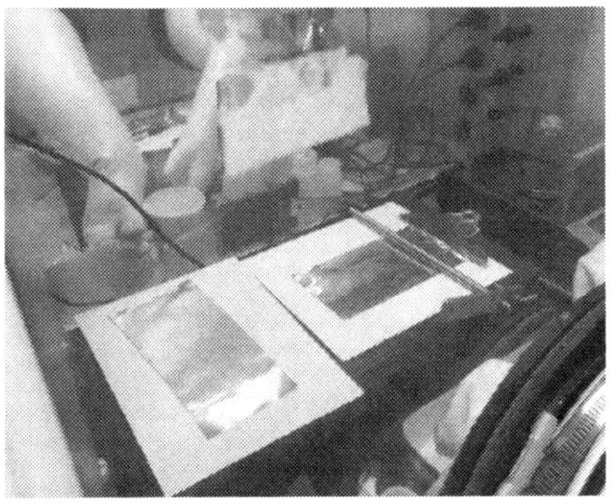

Fig 1. Draw-down machine with Mayer rod set in place on aluminum foil backing.

2.4 Etching away the aluminum and floating targets onto frames

After the foil was baked to set the binder and allowed to cool, pieces of each foil were cut into small 1.5-2 cm squares. The aluminum backing was etched away using a bath containing NaOH (50% w/w). For each foil, this process was 30-40 minutes long. The foil was transferred from the NaOH to a water bath using a microscope slide. The water bath helped to remove remaining NaOH off the target and the targets were floated onto target frames.

2.5 SEM and EDS analysis of the produced foils

Samples were examined by scanning electron microscopy (SEM) with energy dispersive x-ray spectroscopy (EDS) analysis to examine the structure of the printed carbon foils and the elemental composition of the pure, purified, and impure samples. SEM and EDS were performed using a Hitachi S-3000N SEM equipped with a Thermo-Noran UltraDry detector and System Seven analysis system and software. Images were taken at 10 keV using a secondary electron detector. Composition analyses by EDS used a "Proza" (φ(ρz)) standardless analytical correction, assuming metals (i.e. oxygen was analyzed directly rather than assuming oxides).

3 Results and conclusions

3.1 Removal of impurities

A mock ^{14}C stock feed sample was prepared using natural carbon, magnesium oxide, aluminum oxide, and silicon oxide. A simple dissolution method was developed to remove the impurities and minimize the loss of the valuable, enriched ^{14}C material. The oxides present an interesting chemical processing issue as the magnesium oxide is a basic oxide, aluminum oxide is an amphoteric oxide, and silicon oxide is a weakly acidic oxide, so multiple solvents would be necessary to full dissolve away all the impurities. Since NaOH is part of the process already, it was decided to first begin this process using base to dissolve the silicon oxide and potential a small portion of the aluminum oxide (given a sufficient contact time). After the sample was contacted with the NaOH, it was noted that a white oxide layer formed on the surface of the bulk carbon layer. Assumedly, the larger grain size of the carbon enabled it to sink faster than the finer oxides. In one of the samples, this oxide layer was removed to see how this affected the overall retained impurities (Figure 2). During the NaOH dissolution and the first water wash, an oily carbon layer was observed. This layer shrank with each wash and likely resulted from some the interaction of the base and oxides. The water was significantly clearer after the second water wash (Figure 3). These materials transferred using isopropanol to clean polypropylene jars, allowed to dry, and then used to produce carbon paint for preparing foils. Final determination of purity was performed by SEM and EDS, which is reported later in this paper.

Fig. 2. White oxide layer formed on top of the sample (left) and resulting amount of oxide layer after removal (right). Some losses of carbon due to pipetting.

https://doi.org/10.1051/epjconf/202022904002

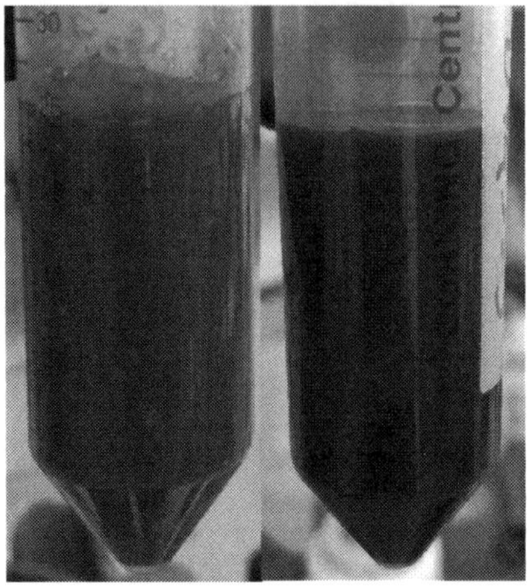

Fig. 3. After the material was combined with NaOH and shaken, note the frothiness and bubbling (Left). After the material was washed a second time with water, the solution is homogenous (right).

3.2 Printing the carbon on an aluminum foil

The carbon powders (natural carbon, purified with oxide layer removed (PWR), purified without oxide layer removed (PWOR), and carbon/impurities mixture) were successfully converted into paint, and foils were prepared for each sample. After the foils were baked, there is a distinct color variance between the various samples (Figure 4). The foils of natural carbon and the PWR sample are similar in color. The PWOR sample is slightly grayer. The unpurified carbon sample is very noticeably grayer than the rest. This suggests the method effectively removed a significant amount of the impurities. Additionally, the brittleness of the target increased with the amount of impurities present (as is evident from the flakiness of the impure foil in Figure 4).

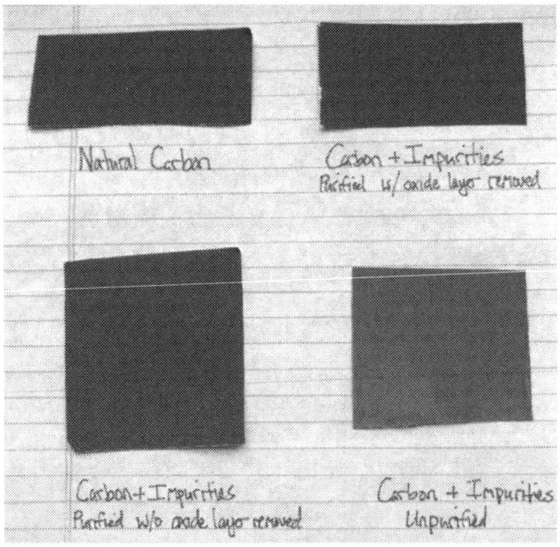

Fig. 4. Foils produced from each carbon sample.

Based on the known areal density of the aluminum foil, the areal density of each foil was determined. Each of the foils were between 2.1-3.2 mg·cm^{-2} thick. All foils survived being etched. A vial was placed inside the etching container to keep the foil upright, so the bubbles did not get trapped under the foil surface and damage the foil. Finally, the foils were floated in water and attached to the target frames.

3.3 SEM and EDS analysis of the produced foils

SEM and EDS were performed on sample foils of the various carbon purities. Purely observing the foils under SEM, the PWR sample appears quite similar to the natural carbon sample, the layer is quite defined and the material looks well packed. Conversely, the impure carbon and the PWOR sample show a pitted less structured formation (Figure 5), which explains the brittleness observed when handling the foils.

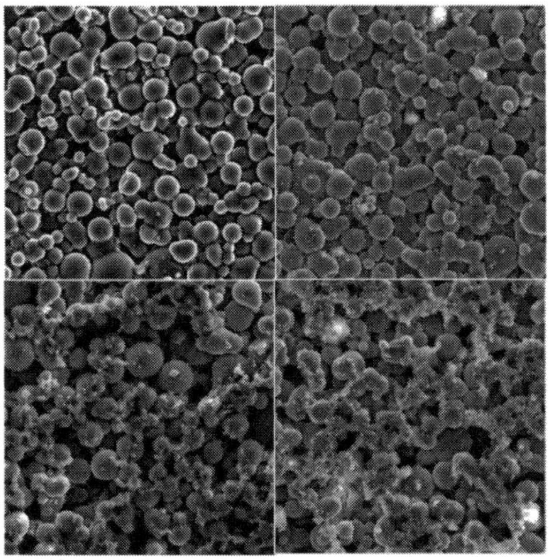

Fig. 5. Comparison of the SEM spectra for all produced foils. Magnification was x1000. Natural Carbon (Top Left), Purified with Oxides Removed (Top Right), Purified without Oxides Removed (Bottom Left), and Impure Carbon (Bottom Right).

EDS was performed on each sample to determine the efficacy of the dissolution method to remove the unwanted impurities. Relative to the impure sample, 99.9% of the silicon dioxide was removed for both methods of oxide removal. This is expected with silicon dioxide being an acidic oxide, it is highly soluble in basic solutions. For the PWOR sample, only a small percentage of magnesium oxide (4.1%) and aluminum oxide (6.5%) were removed. For the PWR sample, 53.9% of the magnesium oxide and 57.3% of the aluminum oxide were removed. Simply pipetting was capable of removing the majority of these impurities, which really lead to a significantly more stable target being produced. Figure 6 shows representative EDS spectra comparing the impure carbon sample versus the PWR sample.

https://doi.org/10.1051/epjconf/202022904002

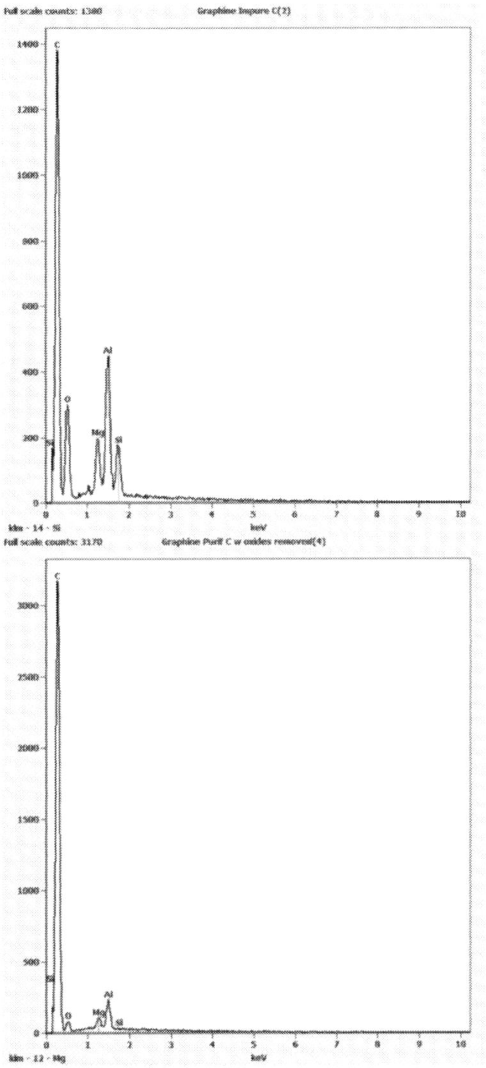

Fig. 6. Comparison of the EDS spectra for the impure carbon (top) and purified carbon with the oxide layer removed (bottom).

4 Conclusions

A simple, glove-box-friendly method was developed to remove the bulk oxide impurities found in the ^{14}C stock material at Argonne National Laboratory. The purified material was utilized in a simple dry down method to produce thin foils for nuclear physics experiments. The produced foils were evaluated by SEM/EDS analysis to examine the structural integrity of the foils and evaluate the reduction of the incorporated oxides. The method successfully produced stable foils with the bulk oxide impurities removed. In the future, additional steps can be added if necessary to improve the dissolution and removal of the magnesium and aluminum oxides.

The authors gratefully thank Jeff Fortner (Argonne) for performing the SEM/EDS analysis. This material is based upon work supported by the U.S. Department of Energy, Office of Science, Office of Nuclear Physics, under Award Number DE-SC-0017208 and Contract Number DE-AC02-06CH11357. This research used resources of Argonne's ATLAS facility, which is a DOE Office of Science User Facility.

References

1. I. Pavlovsky, R.L. Fink, M.D. Gott, and J.P. Greene, AIP Conference Proceedings, **2160**, 040009 (2019)

Target preparation for neutron-induced reaction measurements

Goedele Sibbens*, *Alf* Göök, *David* Lewis, *André* Moens, *Stephan* Oberstedt, *David* Vanleeuw, *Ruud* Wynants, and *Mariavittoria* Zampella

European Commission, Joint Research Centre, Directorate G, Retieseweg 111, B-2440 Geel, Belgium

Abstract. Neutron-induced reaction measurements, e.g. cross-sections and particle emission yields, require samples, called "targets", with specific properties depending on the reaction being studied and the quantities being measured. The target characteristics influence the results of these measurements and can have a strong impact on the total uncertainty of neutron-induced reaction data, which are important for industry and in research. This paper gives an overview of the main techniques applied in the target preparation laboratory at JRC-Geel for production and characterization of targets for neutron-induced reaction data measurements. The use of different targets is demonstrated with a few examples of total and reaction cross-section as well as nuclear fission experiments. In addition, on-going investigations and technical developments are presented.

1 Introduction

High-quality and well-characterized samples, called "targets", are one of the pillars of nuclear data experiments together with the nuclear facilities, the data acquisition systems, and the theoretical and numerical data processing. Nuclear targets consist of a freestanding or a supported material on a substrate and range in thickness from millimetres to nanometres. They have to be physically and chemically stable. The target substrate is as thin as reasonably possible, mechanically stable and should not undergo any nuclear reaction leading to products that interfere with the material under investigation.

The data is generated by measurements at the two large-scale neutron facilities at the JRC-Geel. Both facilities rely on accelerators which provide charged particle beams that in turn produce neutrons through nuclear reactions. At GELINA a 150 MeV linear accelerator produces an electron beam which hits a rotating uranium disc. When the beam hits the uranium disc, Bremsstrahlung is produced, leading to neutrons via (γ,n) and to a lesser extent (γ,f) reactions. At MONNET, a 3.5 MV light-ion Tandem accelerator which replaced recently the 7 MV light-ion Van de Graaff accelerator, provides charged particle beams, which allow producing quasi mono-energetic neutrons up to 20 MeV through reactions between a suitable particle beam and a neutron-producing target [1, 2]. Already from the 60's the CBNM (Central Bureau for Nuclear Measurements), renamed IRMM and later more recently JRC-Geel, was preparing targets for nuclear physicists who were measuring neutron-induced reaction cross-sections as recommended on the priority list of the EANDC (European-American Nuclear Data Committee) which was later reorganised to form the NEA (Nuclear Energy Agency) of the OECD (Organisation for Economic Co-operation and Development) [3, 4]. The main users of these nuclear data are European Union member state organisations and international partners of the EURATOM program, in particular European safety authorities, the nuclear industry, the nuclear research community, the OECD NEA and the IAEA (International Atomic Energy Agency) [1]. A history of the techniques applied at JRC-Geel to prepare and characterize targets is described in [5, 6].

This paper describes the main techniques applied nowadays in the target preparation laboratory at JRC-Geel for the production and characterization of nuclear targets. The use of these targets is demonstrated with a few examples of neutron cross-section measurements. In addition, ongoing investigations related to molecular plating, pellet compacting, e-beam evaporation, cleaning of glass plates for the production of polyimide foils and vacuum deposition parameters are presented.

2 Neutron-induced reactions

Neutron-induced reactions are of importance for various applications, e.g. energy production and medical treatment. For the quantification of the energy production in a nuclear reactor or the applied dose to the tissue in the human body we need to know the cross-sections for the various reactions a neutron may undergo with different elements (isotopes) present in a nuclear reactor or a human body, respectively. Cross-sections and secondary particle emission production yields, e.g. alpha particles, gamma-rays, secondary neutrons and fission fragments, need to be known as a function of the incident neutron energy. Dedicated measurement

* Corresponding author: goedele.sibbens@ec.europa.eu

programmes are being executed at the two JRC neutron sources (GELINA, MONNET) since more than 50 years.

Different types of neutron-induced reactions will lead to different reaction products, e.g. neutrons from elastic and inelastic scattering reactions $^{A}X(n,n)^{A}X$ and $^{A}X(n,n'\gamma)^{A}X$ respectively, photons from radiative capture reactions $^{A}X(n,\gamma)^{A+1}X$ or fission fragments from neutron-induced fission reactions [7]. Other neutron absorption processes can lead to proton or alpha particle emission. Fig. 1 shows a schematic overview of a number of possible neutron-induced reactions.

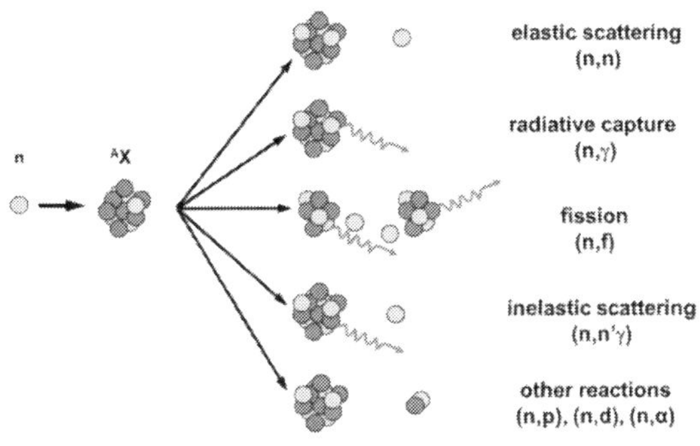

Fig. 1. Neutron-induced reactions.

The neutron cross-section is a measure of the probability for an interaction between an incident neutron and a target nucleus. Neutron cross-section measurements can be divided into transmission measurements and reaction cross-section measurements. In transmission measurements the observed quantity is the fraction of the neutron beam that travels through the target without any interaction. Targets for transmission measurements have to be homogeneous without holes and covering a range of thicknesses. As a general rule, a transmission between 0.2 and 0.7, depending on the background conditions, can be recommended. To ensure the homogeneity of the target, for some elements, measurements are performed on powder mixtures (with sulphur or carbon) or solutions. In reaction cross-section measurements the observed quantity is the yield which is the fraction of the neutron beam that undergoes a reaction in the target. For reaction cross-section measurements in which the escape probability of the reaction product is strongly influenced by the target and/or the substrate thickness, targets are preferably thin homogeneous layers on thin backings [8].

For the understanding of the nuclear fission process the fission fragment production yields, the mass-resolved kinetic energy distribution and the characteristics of the particles, neutrons and gamma-rays, emitted from the highly excited fragments shortly after scission of the fissioning nucleus needs to be known. Usually two different techniques are applied to measure fission fragment emission yields, either recoil mass-spectrometry [9] or the double-energy/double-velocity technique [10-12]. Pre-requisite for precise measurements performed with either of the two techniques is to use very thin and homogenous actinide deposits to minimize the energy loss of the fission fragments when passing that layer. For the double-energy/double-velocity technique it is important that the substrate, onto which the very thin actinide layer is deposited, is thin and homogenous, too. In the ideal case the energy loss should be so small that light and heavy fragments are separated, independent of the path length which the ion has to pass through the target or the target and the substrate layer.

In general for neutron-induced reactions the areal density of the nuclide under investigation needs to be known. In addition, the homogeneity of the target and the presence of contaminants and impurities, which can influence the measurement, need to be checked.

3 Target preparation and characterization of thick targets

3.1 Mechanical transformation

Several mechanical transformation techniques are applied to re-shape material. In the case of radioactive material, the re-shaping is done in a glove box in the nuclear controlled area. For oxidising material an argon glove box is used as far as possible, otherwise the material is protected with an oil or petrol layer from reaction with the environment. A piece of metallic material can be transformed into a thin foil with a thickness between 1.0 mm and 0.05 mm by rolling and, afterwards, into a disc with a diameter ranging from 1 mm to 100 mm and a thickness up to 0.5 mm by punching. Metallic wires can be produced with a diameter down to 0.5 mm by wire drawing.

3.2 Powder compacting

Powder targets are prepared by canning the powder in an Al container in which the powder is slightly manually pressed and the cover of the Al can is glued. In case of pellets the powder is mixed with sulphur powder and cold compacted with a hydraulic press. Sulphur has been chosen as binder for form stability. The required pressure to prepare pellets depends on the diameter of the matrix and the die. In case of radioactive material, the preparation of powder targets is performed in a glove box in the nuclear controlled area.

3.3 Characterization

Metal targets like discs and foils are characterised for their mass and dimensions. The mass is determined by accurate weighing with a microbalance applying the substitution method. In substitution weighing, the mass of the target is determined via the so called "SUUS" method in which a series of mass determinations of an unknown (U) and a reference weight (S) is performed. The uncertainty contributions are the uncertainties associated with the calibrated reference weights, the air buoyancy correction and the repeatability of the balance. The diameter of the targets is measured accurately using a non-contact vision measuring microscope based on

leading-edge optical technologies with a readability of 0.5 μm. The contour of the target is scanned several times at both sides. In case of a powder pellet the diameter is not measured directly, instead the inside diameter of the press tool is taken. The thickness measurements are done with a calliper, in case of powder pellets, or a thickness gauge for the other targets. Before each set of measurements the measurement instrument is calibrated with reference gauge blocks. The uncertainty contributions are the uncertainties coming from the reference gauge blocks and the repeatability of the measurements.
Table 1 gives an overview of present preparation and characterization techniques for thick targets.

Table 1. Overview of the present preparation and characterization techniques for thick targets.

Method\Material	Metal	Powder
Preparation		
Mechanical transformation	Cutting, rolling, punching and drawing equipment	
Powder compacting		Hydraulic press
Characterization		
Weighing	Balance	Balance
Area's and thickness measurement	Vision measuring microscope and thickness gauge	Calliper

4 Target preparation and characterization of thin targets

4.1 Thin foil preparation

For neutron-induced reaction measurements where the energy loss of the reaction product in the substrate needs to be minimized, thin Al foils with a thickness of 25 to 30 μm on an Al ring can be used. Fig. 2 shows the stretching process of the Al foil. The foil is first locked between two rings and then placed on a support for stretching. The Al rings are glued on the stretched Al foil and later cut out.

Fig. 2. Stretching process of an Al foil.

Another type of thin substrate is a polyimide foil with a typical areal density of 35 $\mu g \cdot cm^{-2}$ produced by in-situ polymerization [13]. The first two steps of this method, which are the preparation of the polycondensate solution and the glass coating, are done in an argon glove box to keep the relative humidity lower than 15%. The third step which is the polymerization is performed at 350°C in a dust-free oven. Finally the foils are cut on the glass plate and transferred on a frame via a water surface in a laminar flow cabinet. The polyimide foil can be coated afterwards with a 35 $\mu g \cdot cm^{-2}$ Au layer by physical vapour deposition to become electrically conductive.

4.2 Thin foil characterization

The areal density of the polyimide foils is measured with a visible light spectrophotometer. With this instrument the amount of light is measured that is refracted and/or reflected from the top and the bottom of the polyimide foil as function of the wavelength. The interference pattern is used to calculate the foil thickness. When the foil is still on the glass plate, the light reflection mode is applied. After transferring the foil on a frame, the areal density is re-measured via the light transmission mode [13]. The relative uncertainty on the areal density is in the order of 10%.

4.3 Thin film deposition

Thin film deposits with an areal density ranging from 10 to 300 $\mu g \cdot cm^{-2}$ are prepared by two kinds of techniques: physical vapour deposition and the so-called "molecular plating". In case the ingrowth of a daughter nuclide would interfere with the characterization and neutron-induced reaction measurements, the actinide material is purified prior to the deposition on a substrate. The radiochemical separation is performed by ion exchange [14].

4.3.1 Physical vapour deposition

$^{235,238}U$, Li, B, tristearin and Au layers are prepared by physical vapour deposition in different set-ups, dedicated to each material to avoid cross-contamination. The principle of this system is to evaporate or sublimate the starting material in a vacuum and condense the vapour on a substrate.

The evaporators, except for B, consist of a Ta crucible connected to a power supply in a closed chamber, connected to a pump system in order to work under vacuum conditions as described in [16]. The distance between the material to be evaporated and the substrates is determined according to the traveling distance of the evaporated atom or molecule, the required homogeneity and yield and is set at 195 mm in the evaporators for $^{235,238}U$, at 400 mm for Li, at 200 mm for tristearin and at 250 mm for Au. Al rings are mounted as masks in front of the substrates and determine the requested area of the deposited material. The deposit thickness is monitored with a quartz crystal microbalance during the deposition process.

For the evaporation of ^{10}B, which has a melting point of 2349 K (2076°C), a bent electron beam (e-beam) bombardment source is used for the heating of the B material. In this case the evaporation rate is controlled by the e-beam current. The old e-beam evaporator for B has been replaced by a new one which has multiple water cooled cupper pockets with a tungsten liner for the source material. The deposition rate is monitored with a quartz crystal microbalance. The distance between the evaporation source and the substrate can be changed from 300 to 400 mm. Prior to a deposition the substrate can be plasma cleaned with a glow discharge module. The equipment has the possibility to heat the substrate during and after deposition. More information of the new e-beam evaporator including B evaporation tests and calibration of the thickness monitor by weighing and neutron time of flight measurements has been described by Vanleeuw et al. in [15].

To prepare deposits of ^{235}U [16] and ^{238}U [17] by resistance heating in vacuum, the U_3O_8 starting material needs first to be converted into UF_4 by a wet chemical precipitation method because U_3O_8 will be reduced into more stable $UO_{2\pm x}$ when heated in vacuum and so temperatures much higher than the U_3O_8 melting point of 1423 K are needed for the deposition process which is not the case for UF_4. Each uranium evaporator is integrated in a glove box in the nuclear controlled area.

The vacuum deposition of metallic Li is performed in a multi-crucible evaporation set-up that consists of a tantalum crucible heater surrounding an alumina crucible for the Li material. A second crucible of tantalum is present to be able to evaporate a 50 $\mu g \cdot cm^{-2}$ Au passivation layer on the substrate prior to the Li deposition and a 50 $\mu g \cdot cm^{-2}$ Au protective layer coating the lithium deposit directly after evaporation to avoid that Li reacts with the substrate and the environmental elements, respectively [18].

The above mentioned techniques are applied one after another in case of a multi-layer target [19].

4.3.2 Molecular plating

The molecular plating technique is based on the cathodic deposition of the actinide material onto an Al substrate in isopropanol at a constant voltage of 360 V. For each isotope of U, Pu, Am and Np a molecular plating cell is designed and produced in the JRC-Geel workshop according to the required dimensions of the substrate and the deposit. The anode is a platinum grid rotating at a speed of 5 to 10 rounds per minute. The cathode is connected to a stainless steel holder which contains the Al substrate with an Al ring as mask that determines the diameter of the actinide deposit. Prior to the deposition the Al substrate is chemically etched in a mixture of 80% H_3PO_4 + 4% HNO_3 + 16% H_2O. The original oxide material is first dissolved in 14 M HNO_3, evaporated to near dryness and subsequently re-dissolved in 0.75 M HNO_3 at a concentration linked to the required plated amount on the substrate. Volumes of 20 up to 100 µL of the actinide nitrate solution are added to 20 mL up to 200 mL of isopropanol depending on the volume of the cell. After two hours of molecular plating the Al substrate with deposit is removed from the cell and heated up on a hotplate for 5 minutes at 120°C to evaporate the remaining isopropanol [20, 21]. To avoid that dust can enter via the top part and to re-load and empty the cell in a faster and safer way, the upper part has recently been redesigned. Fig. 3 shows pictures of the new molecular plating cell.

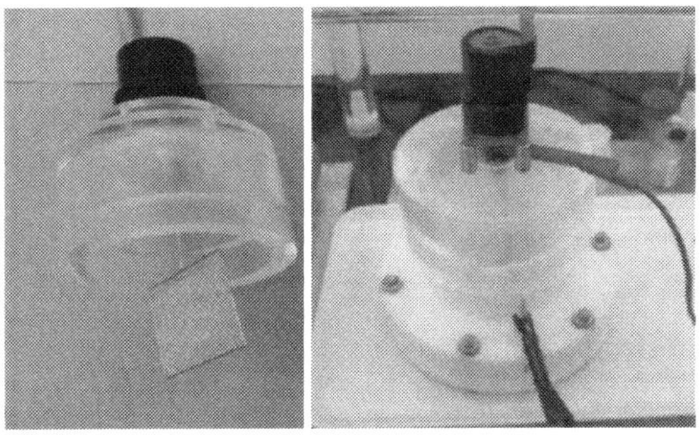

Fig. 3. Pictures of the molecular plating cell; left part: new upper part with Pt anode, right part: cell connected to power supply.

4.4 Characterization

The total activity of the actinide layers is determined by low-geometry alpha-particle counting [22]. The measurements are carried out using two different set-ups: an alpha-particle counter with variable (50-200 mm) and with fixed (1637 mm) target-to-detector distance as described in [14]. The solid angle is determined by the distance of the target to the diaphragm in front of the detector, the diameter of the diaphragm and of the deposited layer. The latter is not measured directly; instead the inner diameter of the mask, used during the physical vapour deposition or molecular plating, is measured. For the alpha counter with variable length, the distance from the target to the diaphragm in front of the detector is measured with calibrated gauge blocks. From the measured activity, the isotopic composition and deposited area, the mass and areal density of the isotope of interest is calculated. The uncertainty contributions are the uncertainties associated with the counting statistics, the corrections for background, for decay, for ingrowth of daughter nuclide, the measurement time, the isotopic composition and the solid angle.

A morphological and compositional study of ^{238}U thin film targets showed that layers prepared by vacuum deposition are smooth UF_4 deposits [21]. Results of homogeneity measurements showed a variation of the areal density of the order of 5 % between the centre and the edge of deposits with a diameter of 70 mm [22]. The targets prepared by molecular plating are maze-like layers of U(VI) co-deposited with isopropanol [21].

The mass of stable deposits is determined by difference weighing. Therefore, the mass of the substrate is measured on a microbalance before and after deposition by means of the substitution method. The areal density is calculated from the mass, corrected for isotopic abundance and impurities, and the diameter of the deposit that is determined by measurement of the inner diameter of the mask used during the physical vapour deposition. The uncertainty contributions are the uncertainties associated with the calibrated reference

Table 2. Overview of the present preparation and characterization techniques for thin targets.

Method\Material	^{235}U, ^{238}U	U, Pu, Am, Np	LiF Metallic Li	$C_{57}H_{110}O_6$	Au	^{10}B
Preparation						
Vacuum deposition	Resistance		Resistance	Resistance	Resistance	e-beam
Molecular plating	Isopropanol	Isopropanol				
Characterization						
Weighing			Balance	Balance	Balance	Balance
Area's measurement	Calliper	Calliper	Calliper	Calliper	Calliper	Calliper
α-particle counting	Low geometry	Low geometry				
Combustion analysis				in O_2		

weights, the air buoyancy correction, the repeatability of balance, the isotopic composition, the impurities, the reference gauge blocks and the repeatability of the diameter measurements.

In case of very thin boron deposits which cannot be characterized by difference weighing the areal density is determined by neutron time of flight measurements [15].

For tristearin deposits the hydrogen content is determined by combustion analysis according to the principle of catalytic tube combustion in an oxygenated atmosphere [23].

An overview of the techniques for the preparation and characterization of thin targets for neutron-induced reaction experiments is shown in Table 2.

5 Examples of neutron-induced reactions

5.1 Neutron fluence monitors

In neutron reaction cross-section measurements, the neutron fluence needs to be accurately determined. This neutron fluence is measured through the detection of charged particles that are produced in standard neutron-induced reactions. Depending on the neutron energy region of interest, a specific standard reaction is used. In the thermal and resonance energy region, ^3He, ^6Li and ^{10}B are used to determine the neutron fluence in the beam. In the case of ^6Li and ^{10}B, the method is based on the detection of alpha particles produced in the (n,α) reactions $^6Li + n \rightarrow {}^3H + \alpha$ and $^{10}B + n \rightarrow {}^7Li + \alpha$.

Fig. 4. Multi-layer ^{235}U based fluence monitor.

In the fast neutron energy region, fluence measurements are based on the detection of fission fragments produced in the standard ^{235}U(n,f) reaction. Typical detectors used for fluence measurements are gas-filled (e.g. ionisation chamber) and solid-state detectors or scintillators (e.g. Li-glass detector). In both cases, thin film layers are needed to limit energy loss of the reaction products within the layer. Fig. 4 shows the interior of a multi-layer ^{235}U based fluence monitor.

In the high-energy range, for mono-energetic neutrons up to 24 MeV, a recoil proton telescope can be used to measure the neutron fluence. It makes use of the (n,p) cross-section. The instrument consists of a tristearin ($C_{57}H_{110}O_6$) target as radiator to generate the protons and a silicon surface barrier detector as solid angle proton counter. The number of detected recoil protons is directly linked to the areal density of hydrogen in the tristearin target. The hydrogen content of the tristearin bulk and deposited material has been determined by combustion analysis. The stability of tristearin has been demonstrated even for long evaporation runs and a reliable value of the hydrogen content of 12.35 (10) wt% has been established [23].

5.2 Neutron-induced cross-section experiments

For ^{233}U(n,n'γ) cross-section measurements at GELINA [24] a 8.3 g ^{233}U disc with a diameter of 30 mm, a thickness of 0.64 mm and an activity of 3 GBq has been prepared by punching (left part of Fig. 5). The (n,xnγ) cross-sections of actinides involved in new generation cycles like the thorium cycle, are important to test and improve predictive power of theoretical codes allowing a better knowledge of the neutron inelastic scattering process. The ^{233}U disc was prepared by punching a 76.2 x 38.1 x 0.64 mm U foil in a glove box in the nuclear controlled area. Because of the limited ^{233}U material simulations have been done with Al, stainless steel, Ta and U to find the optimum punching parameters. Finally the ^{233}U material bagged in a self-sealing bag and a sealed single PVC bag, was punched with the P/O Weber 300 kN press at a force of 20-25 kN in combination with a die of 30 mm diameter. After weighing the ^{233}U disc, the target was transferred to an argon box for mounting between two Al foils with a thickness of 50 μm each in a tailor-made holder that maintains the Ar atmosphere and allows placing it in a neutron beam of at most 55 mm diameter with minimal support material in the beam. The diameter of the U disc has not been measured; instead the diameter of the punch die has been determined by measuring the diameter of two punched test samples.

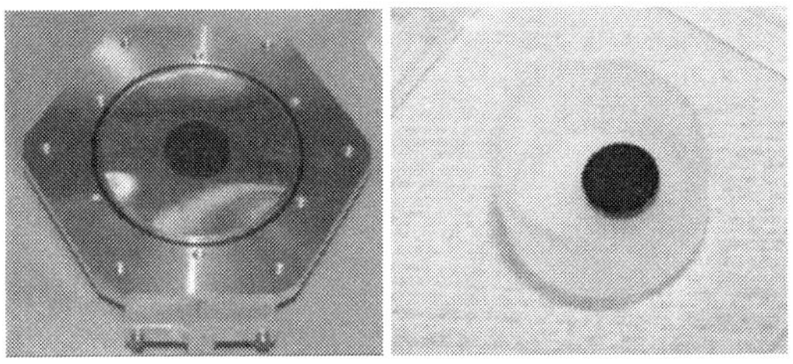

Fig. 5. Left part: ^{233}U disc with a diameter of 30 mm and thickness of 0.64 mm prepared by punching and mounted in a measurement holder with 50 µm Al foil. Right part: Picture of a 5 mm diameter pellet of 0.1 g $^{238}U_3O_8$ in a plastic container.

A $^{238}U_3O_8$ pellet (right part of Fig. 5) with a mass of 0.1g, a diameter of 5 mm and with a low $^{236}U/^{238}U$ ratio of $7 \cdot 10^{-12}$ has been prepared with a Specac 15 Ton hydraulic press operating at 10 kN and a 5 mm diameter die. The mass has been determined by weighing. The pellet is used in a campaign to measure the $^{235}U(n,\gamma)$ cross-section. The cross-section will be deduced by quantifying the number of ^{236}U nuclei produced after neutron irradiation at FRM II (Forschungs-Neutronenquelle Heinz Maier-Leibnitz) at the Technical University of Munich.

To investigate the mass, kinetic energy and nubar variations with the resonance energy and spin in the resolved resonance region of the ^{239}Pu fission cross-section, a ^{239}Pu layer with a diameter of 50 mm and an areal density of 29.95 $\mu g \cdot cm^{-2}$ was prepared by vacuum evaporation on a polyimide foil of 38 $\mu g \cdot cm^{-2}$. The importance of an excellent target to achieve an optimum mass and energy resolution over all emission angles is demonstrated in Fig. 6. The figure presents the fission fragment pulse height versus the cosine of the emission angle depicted for fission fragments detected in $^{239}Pu(n,f)$ reactions obtained in a double-energy measurement with a twin Frisch-grid ionization chamber. The left part shows fragments emitted to the layer side, and the right part fragments emitted through the substrate in opposite direction. The distribution on the left side of Fig. 6 reflects the nearly perfect homogeneity of the active layer by a complete separation of light and heavy fragments irrespective of the emission angle. Also, the distribution obtained after passing the polyimide foil (right part of Fig. 6) shows still very good separation down to $cos\theta = 0.3$, which corresponds to an emission angle larger than 70° relative to the beam axis. This high-quality pulse-height distribution allows obtaining fission fragment yield distributions with a mass resolution better than 5 mass units at FWHM. It is thanks to an excellent thin-layer target that fission cross-section and fragment yield measurements may be performed with a geometrical efficiency close to 2π.

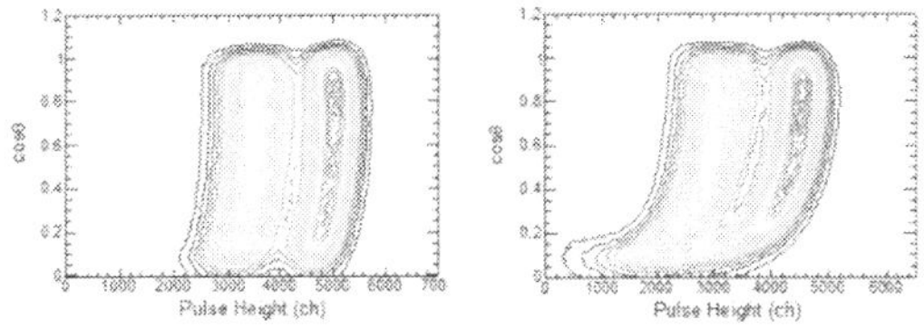

Fig. 6. Fission fragment pulse height versus the cosine of the fission axis relative to the axis of the incident neutron beam as obtained from the reaction $^{239}Pu(n,f)$. Left part: pulse height after having passed the target layer, right part: pulse height after passing the target layer and the polyimide substrate.

6. Investigations on-going

6.1 Molecular plating on Al backings with a thickness smaller than 1 µm

The possibility to deposit thin layers of material on extremely thin backings by molecular plating is being investigated. The thin backings used for this purpose are made of Al foils of 6, 2 and 0.8 µm thickness mounted on Al rings of 0.5 mm thickness, 30 mm inner diameter and 40 mm outer diameter. The molecular plating cell has been re-designed to reduce the mechanical stress on the Al foil and minimize the occurrence of foil-break during the mounting and disassembling of the cell and during the molecular plating process.

6.2 Production of recyclable powder pellets

Investigation is on-going for the production of recyclable uranium and plutonium powder pellets prepared by cold compacting. In case of limited and valuable material, it is favourable to be able to recuperate the powder afterwards so that it can be used for other experiments. In this case mixing the powder with a binder like sulphur powder is not recommended as purification of the original powder can become complicated and time consuming especially in case of highly radioactive material. Tests are ongoing to find the optimum press parameters for different diameters and amount of material to produce mechanically stable U_3O_8 pellets and to set the boundaries of the preparation process.

6.3 Cleaning of glass plates for polyimide foils

Cleaning of glass plates used in the production process of polyimide foils is still time consuming and it is difficult to get the plates hydrophilic, i.e. the spreading of water on the surface of the glass plate should be uniform. Investigation is on-going to use oxygen plasma to "soft" etch the surface instead of cleaning the glass plates by rubbing with a paper towel and water as reported in ref. [13]. Several glass plates are cleaned in the plasma etching system under different cleaning conditions. They are covered with the polyimide solution and placed in the oven at 350°C for polymerization. The ability to release the foil from the glass plate is related to the surface of the glass plate. Tests are being conducted to find the optimum parameters for the plasma pod.

6.4 Preparation of uranium deposits by e-beam evaporation

In the frame of the replacement of the old ^{235}U evaporator, an investigation is on-going to prepare thin uranium layers by electron beam physical vapour deposition instead of evaporative deposition in which the material to be deposited is heated to a high vapour pressure by electrically resistive heating. The advantage would be that the original U_3O_8 material can be used instead of first converting this material into UF_4 by a time consuming wet chemical precipitation method.

6.5 Production of an evaporator

Commercially available vacuum evaporators are not suitable to explore vacuum deposition parameters. A design for an evaporator has been made and the equipment, dedicated to stable isotopes, will be produced and used to better understand the vacuum deposition process.

7. Conclusion

For neutron-induced reaction measurements targets are produced according to the reaction being studied. Thick targets like discs and foils are prepared by cutting, rolling and punching and by cold compacting powder pellets. The targets are characterised for their mass by accurate weighing and for the area by accurate vision measuring microscopy. Thin deposits on a substrate are prepared by physical vapour deposition and molecular plating. The area of the deposits is determined by measuring the inner diameter of the mask during the deposition process. The mass of the stable deposits is determined by accurate weighing and of the actinide deposits via the activity measured by low-geometry alpha-particle counting.

The authors thank the JRC-Geel colleagues: the workshop for the production of the molecular plating cells, containers, masks etc., the nuclear chemistry and nuclear mass spectrometry group for the isotopic composition of the actinide material, Jan Heyse for the characterization of boron targets by n_TOF measurements and the engineering materials laboratory for the SEM-EDX spectroscopy of stable targets; and the JRC-Karlsruhe colleagues: the analytical service for the ICP-MS analysis, the surface science laboratory for the XPS and the electron microscopy group for the morphology and elemental composition analysis by SEM-EDX.

References

1. https://ec.europa.eu/jrc/en/research-topic/neutron-data
2. W. Mondelaers, et al., Il Nuovo Cimento **38** C, 175 (2015)
3. https://www.oecd-nea.org/science/nucleardata.html
4. http://www.oecd-nea.org/dbdata/hprl/
5. A. Stolarz, et al., Nucl. Instrum. Methods A **613**, 351-356 (2010)
6. G. Sibbens, et al., in *Proceedings of the Scientific Workshop on Neutron Measurements, Theory and Applications, Nuclear data for Sustainable Nuclear Energy, 28 – 30 April, 2009, Geel Belgium*
7. R. Nolte, in *Lecture notes at the Ecole Joliot Curie 2014 "Neutrons and Nuclei, 28.09-3.10.2014, Frejus, France*
8. P. Schillebeeckx, et al., Nucl. Instrum. Methods A **613**, 378-385 (2010)
9. A. Bail, et al., in *AIP Conference Proceedings 1175, 2009, 4th International Workshop on Nuclear Fission and Fission-Product Spectroscopy, Cadarache, France*
10. A. Göök, et al., Nucl. Instrum. Methods A **830**, 366-374 (2016)
11. S. Oberstedt, et al., Phys. Rev. C **90**, 024618 (2014)
12. M.O. Frégeau, et al., Nucl. Instrum. Methods A **817**, 35-41 (2016)
13. G. Sibbens et al., Nucl. Instrum. Methods A **655**, 47–52 (2011)
14. G. Sibbens et al., J. Radioanal. Nucl. Chem. **299**, 1093-1098 (2014)
15. D. Vanleeuw, et al., Preparation and characterization of ^{10}B targets at JRC-Geel, *these proceedings.*
16. G. Sibbens et al., J. Radioanal. Nucl. Chem. **305**, 723-726 (2015)
17. D. Vanleeuw, et al. in *AIP Conference Proceedings 1962, 030008 (2018)*
18. D. Vanleeuw et al., J. Radioanal. Nucl. Chem. **299**, 1113–1120 (2014)
19. G. Sibbens et al., Appl. Rad. Isot. **87**, 229-232 (2014)
20. G. Sibbens, et al., EPJ Web of Conferences **146**, 04062, (2017)
21. G. Sibbens, et al., in *AIP Conference Proceedings 1962, 030007 (2018)*
22. J. Heyse et al., J. Radioanal. Nucl. Chem. **299**, 1055–1059 (2014)
23. D. Vanleeuw et al., J. Radioanal. Nucl. Chem. **305**, 957-962 (2015)
24. M. Kerveno et al., EPJ Web of Conferences **146**, 11012 (2017)

"Thermal Spike" model applied to thin targets irradiated with swift heavy ion beams at few MeV/u

Christelle Stodel[1,*], *Marcel* Toulemonde[2], *Christoph* Fransen[3], *Bertrand* Jacquot[1], *Emmanuel* Clément[1], *Georges* Frémont[1], *Matthieu* Michel[1] and *Christian* Dufour[4]

[1]Grand Accélérateur National d'Ions Lourds (GANIL), CNRS/CEA, Bvd Henri Becquerel, BP 55027, 14076 Caen CEDEX 5, France
[2]Centre de Recherche sur les Ions, les Matériaux et la Photonique (CIMAP), Bvd Henri Becquerel, BP 5133, 14070 Caen CEDEX 4, France
[3]Institut für Kernphysik, Universität zu Köln, Zülpicher Strasse 77, 50937 Köln, Germany
[4]Centre de Recherche sur les Ions, les Matériaux et la Photonique (CIMAP), 6 Bvd du Maréchal Juin, 14050 Caen CEDEX 4, France

Abstract. High electronic excitations in radiation of metallic targets with swift heavy ion beams at the coulomb barrier play a dominant role in the damaging processes of some metals. The inelastic thermal spike model was developed to describe tracks in materials and is applied in this paper to some systems beams/targets employed recently in some nuclear physics experiments. Taking into account the experimental conditions and the approved electron-phonon coupling factors, the results of the calculation enable to interpret the observation of the fast deformation of some targets.

1 Introduction

In nuclear physics, structural evolution of nuclei when moving (far) away from magic numbers [1] remains intriguing in many aspects. It is not well theoretically reproduced so far by state-of-the-art models. High resolution spectroscopic data reveal this nuclear structure evolution as a function of spin, angular momentum, excitation energy and isospin. Their comparison with advanced nuclear model allow a microscopically description of these evolutions.

High resolution gamma-ray spectroscopy sheds light to the nuclear structure by measuring energy of excited states, their decay branching ratios and their lifetime. The Recoil Doppler Distance-Shift (RDDS) technique is an outstanding method to extract the lifetimes of the states of interest to deduce absolute transition strengths that are essential to test theoretical approaches. Here we discuss target problems in measurements using a so-called plunger device for half-life measurements using the RDDS technique [2] combined with the VAMOS++ spectrometer and the tracking germanium detector array AGATA (Fig. 1).

In these recent studies, exotic (i.e. neutron deficient or neutron rich) nuclei of interest were produced in inverse kinematics either by deep inelastic or by fusion-fission and fusion-evaporation reaction using heavy ion beams around the Coulomb barrier.

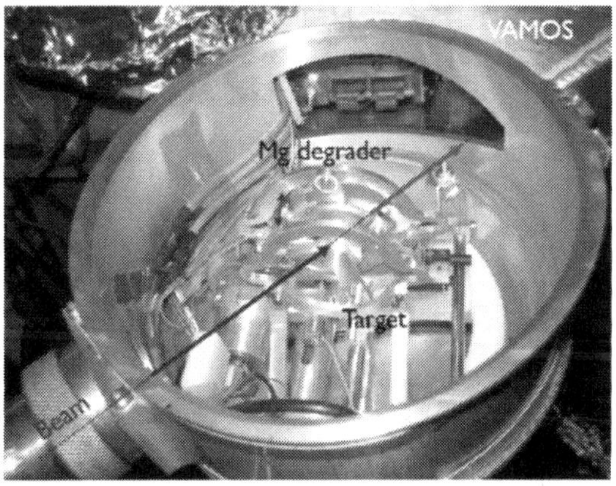

Fig. 1. The Plunger set-up in the AGATA-VAMOS vacuum chamber.

Contrary to previous experiments using a similar set-up but exploring other nuclei produced in direct kinematics, it was observed that some targets were slightly or seriously deformed even at very low heavy ion beam current (< 1 pnA). Unfortunately, this shape change is at the detriment of the quality of measurements: the distance between the plunger target and a plunger degrader foil used to slow down the recoils should be known and constant with a precision of better than 1 μm but the wrinkles height of the material was observed to be of the order of 100 μm, see Fig. 2.

* Corresponding author: stodel@ganil.fr

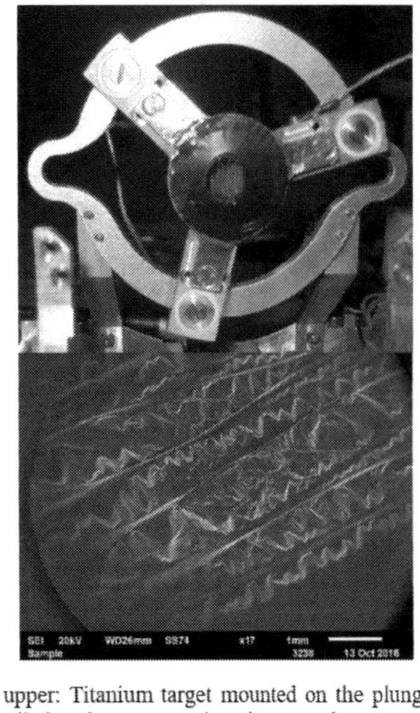

Fig. 2. upper: Titanium target mounted on the plunger set-up after irradiation; lower: scanning electron microscope (electron 20 kV) photograph of the target with magnification *17.

Several causes were then questioned in order to explain the observations. Indeed, the damage can originate from:

- Mechanical stress induced by the fabrication process or the method of stretching and fixation on the frame.
- Chemical reaction with air in the time between fabrication and irradiation under vacuum, some material can oxidize easily and become fragile.
- Interaction with another material in the case of a backing.
- Irradiation inducing heating, phase change or damage of the material.

The last three causes can modify the properties of the material to unexpected values of some parameters (elasticity, conductivity, melting temperature, …) which may make the target delicate to sustain the beams.

In plunger systems, the targets undergo exceptional mechanical strain because they have to be stretched and fixed over a small solid cone. This requirement was verified and qualified before irradiation. But as observed on Fig. 2, the rolling method used to manufacture the target might have induced during irradiation a preferential direction of the wrinkle-like structures of the Ti targets.

In all cases studied, the target temperature at thermodynamically equilibrium were calculated by taking into account the properties of the beam (energy, intensity, spot size) and the target material (emissivity, thermal conductance, thickness). This temperature was estimated considering cooling by radiation and conduction. The values obtained were lower than the fusion temperature of the material reported in literature.

Then the damage formation resulting from a nanometric, dense and shortly created electronic excitation in the material, was investigated using the Inelastic Thermal Spike model (i-TS) [3-6].

Indeed, the passage of a fast ion through a foil creates a "heat spike" along its path, in which the temperature may be briefly extremely high, even if the average temperature of the foil remains low. Within this model, this paper reports on the temperature distribution in time along the beam track calculated for the different projectile/target systems under consideration. The results of the model are compared to the temperature needed for the material to reach the melt phase when the track formation appears in the material.

2 "Thermal Spike" model

2.1 Description

The i-TS model is one of the most powerful tool to describe damage induced by swift heavy ions and is most widely applied to any target material either metallic, semi-conductor or insulators.

The model reproduces the electronic excitation of the material through a four step process: first, the incident ions transfer their energy to the electrons of the target by ion-electron collisions. Second, electron-electron collisions share this energy with other cold electrons. Third, the energy is transferred to the lattice by electron-phonon coupling. Forth, the energy dissipates among the atoms. This process induces a spike along the ion trajectory. Regarding the timeline, illustrated on figure 2.1 from [7], the energy is deposited in the electrons within 10^{-16} to 10^{-15} s and then transferred to the lattice atoms within 10^{-13} to 10^{-11} s.

Mathematically this model is based on specific solutions of two coupled differential equations (1) and (2) governing the heat diffusion to the electrons and atomic subsystems versus time t and space r (in cylindrical geometry) and their exchange via the electron-phonon coupling:

$$C_e(T_e)\frac{\partial T_e}{\partial t} = \frac{1}{r}\frac{\partial\left[rK_e(T_e)\frac{\partial T_e}{\partial r}\right]}{\partial r} - g(T_e - T_a) + A(r) \quad (1)$$

$$C_a(T_a)\frac{\partial T_a}{\partial t} = \frac{1}{r}\frac{\partial\left[rK_a(T_a)\frac{\partial T_a}{\partial r}\right]}{\partial r} + g(T_e - T_a) \quad (2)$$

In these equations, the parameters $T_{e,a}(r,t)$, $C_{e,a}(r,t)$ and $K_{e,a}(r,t)$ are the temperature, specific heat and thermal conductivity of the electronic (e) and atomic (a) subsystem, respectively. $A(r)$ is the energy deposited into the electronic subsystem supplied by the incident ion by ballistic collisions at radius r. The integration of $A(r)$ over space gives the total stopping power dE/dx. The only free parameter in this model is the electron-phonon coupling strength g.

Due to the very short time of energy deposition (10^{-13} – 10^{-12} s), the model calculations are performed within a superheating scenario [8], i.e. that increase of temperature does not stop when reaching the melting or the boiling temperature.

The track radii are associated with the cylinder in which the energy deposited on the atoms surpasses a specific energy. This limit, E_{SH}, is defined as the energy to reach the melting temperature, E_{fus}, plus the energy required to make the solid-liquid phase change (latent heat of fusion, E_{lat}). This criterium is used since the superheating scenario of the target is the result of a very rapid heating stage ($\approx 10^{-13}$ s) [7]. As heat capacity of a material is a function of its internal energy and temperature (equation (3)), the T_{SH} temperature is deduced from the integration of the specific heat from $T_0 = 0\ K$ to T_{SH} which equals the E_{SH} energy.

$$C_a = \frac{dE}{dT} \quad (3)$$

2.2 Application to experiments

2.2.1 Experimental conditions

The experiments were performed at GANIL using heavy ion beams around Coulomb barrier energy impinging on targets as listed in Table 1.

Cases 1, 2 and 3 correspond to experiments using the plunger device mounted in the VAMOS target chamber at an angle of 45° with respect the beam axis. A magnesium degrader slows down the recoiling nuclei produced in the first target placed at an adjustable distance downstream. The first targets were self-supporting and manufactured by rolling. In case 2, a layer of copper was evaporated on the titanium in order to increase its thermal conductivity. The copper layer was facing the beam.

These systems are compared to cases 4 and 5 using either a similar target (Ni) in direct kinematics with a "light" ion beam (^{50}Cr) or a target of ^{54}Fe irradiated with a heavy ion ^{124}Xe beam at comparable velocities.

In case 4, a gold foil was used to slow down the beam before reacting in the nickel target.

In case 5, iron was evaporated on a thin carbon layer, four targets were irradiated and revealed various behaviour.

The effective material thicknesses are given taking into account the angle of the target with respect to the incoming beam (45° in cases 1, 2 and 3 and 0° for cases 4 and 5).

Table 1. Experimental properties of the beam (energy and intensity) and targets (effective thickness).

Experiment 1: $^{238}U^{32+}$(6.2 MeV/u; 1 pnA) + ^{64}Ni (2.6 mg/cm²) + ^{nat}Mg (5.2 mg/cm²)
No problem of target, Mg degrader slightly deformed
Experiment 2: $^{238}U^{32+}$ (6.76 MeV/u; 0.1 pnA) + Cu-^{50}Ti (0.6 - 2.1 mg/cm²) + ^{nat}Mg (4.5 mg/cm²)
Target seriously deformed quickly I_{max}= 3 nAe = 0.1 pnA
Experiment 3: $^{238}U^{32+}$ (6.76 MeV/u; 0.07pnA) + ^{50}Ti (2.1 mg/cm²) + ^{nat}Mg (4.5 mg/cm²)
Target seriously deformed quickly I_{max}= 2 nAe = 0.07 pnA
Experiment 4: $^{50}Cr^{8+}$(4 MeV/u; 10 pnA) + Au (2.4 mg/cm²) + ^{58}Ni (7-10 mg/cm²)
No problem at 30 nAe
Experiment 5: $^{124}Xe^{39+}$(4 MeV/u; 1 pnA) + C -^{54}Fe (0.03 - 0.2 mg/cm²)
3 targets have broken below 6 nAe and 1 sustained up to 80 nAe

2.2.2 Inputs of the i-TS calculations

The initial temperature at equilibrium (T_{eq} in Table 2) of the target is determined by the beam intensity, energy and spot size. It was calculated considering cooling by radiation and conduction through the target holder. The external surfaces radiate towards a black body at room temperature and if necessary, the mutual radiation between two closed targets were considered. The range of temperature is estimated with a beam spot represented by a Gaussian distribution with σ = 1mm and emissivity values ranging from 0.1 to 0.9 [9-10].

The measured thermal conductivity (K_s) of a metal corresponds to the electron conductivity taking into account all the properties of the electron gas. Consequently, using K_s as the input of the metal thermal conductivity, the value of the g factor should be near to the one calculated with a number of electrons equal to the number of atoms [5].

The value of g was determined by fitting the track size confirming the validity of this hypothesis, apart for Ti for which the g factor is twice larger than the expected one. Then the electron-phonon coupling factors g are from [5] and were confirmed or adjusted from detailed study of experimental sputtering yields in nickel [6], titanium [11], and iron [11] materials.

Table 2. Parameters of the materials for the i-TS simulations.

Material experiment	**dP** *[mwatts]*	**T_{eq}** *[K]*	**DE/dx** *[keV/Å]*
Mg			
1	540	850±100	1.9
2/3	48	500±100	1.9
Ni			
1	200	850±100	7.8
4	1700	1150±100	1.5
Cu			
2	4	500±100	7.2
Ti			
2/3	18	450±100	4
Au			
4	250	750±100	2
Fe			
5	10	450±50	4

Material	**g*10¹⁰** **[W/cm³/K]**	**n * 10²²** *[e/cm³]*
Mg	16.8 [5]	4.3 [5]
Ni	107 [5-6]	9.14 [5-6]
Cu	12.7 [5]	8.5 [5]
Ti	1000 [11]	5.7 [5]
Au	2.3 [5]	5.9 [5]
Fe	119 [11, 5]	8.9 [5]

The thermodynamically temperature at equilibrium was well below the fusion temperature of the material for all experiments except for the first one (Mg) where it could have been close.

2.2.3 Results

For comparison, the simulations with i-TS model were done at 300 K and some temperatures around T_{eq} reported in Table 2 in all cases.

The results of the calculations are given in Fig. 3 to Fig. 8 where the lattice temperatures are plotted versus time along ion path. The temperatures are represented by full, dashed and dashed-dotted lines for the initial temperature of the material of 300 K, mean T_{eq} (written in bold and underlined) and $T_{eq} \pm 100$ respectively.

The limit temperature (T_{SH}) corresponds to the energy necessary to melt the material defined as the melting temperature plus the corresponding latent heat of fusion, their values are reported in Table 3 and plotted in the figures where they could be reached together with the corresponding fusion temperature.

Table 3. "Super-Heating" and fusion temperatures [K] of materials.

	Mg	Ni	Cu	Ti	Au	Fe
T_{SH}	1260	2250	1900	2580	1790	2180
T_{fus}	923	1728	1358	1941	1337	1811

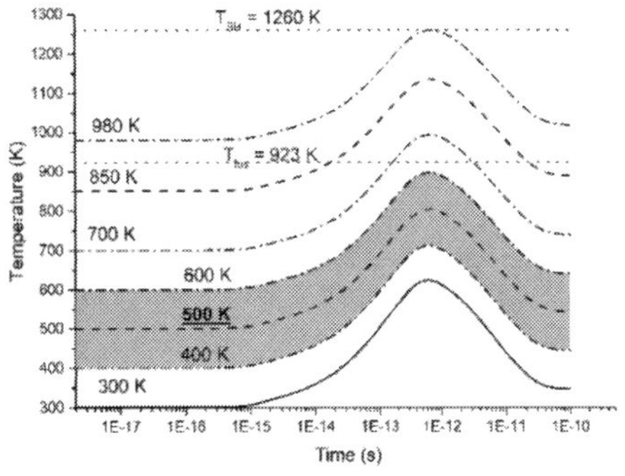

Fig. 3. Evolution of the lattice temperature of the Mg degrader as a function of time along the ion path for cases 1, 2 and 3 (filled in grey).

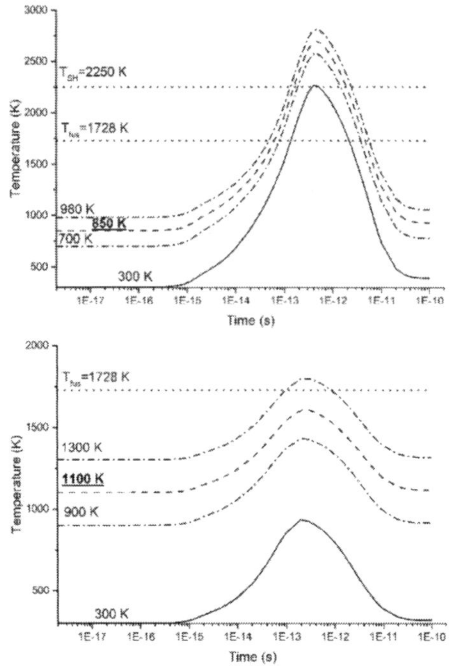

Fig. 4. Evolution of the lattice temperature of the Ni target as a function of time along the ion path for cases 1 (upper) and 4 (lower).

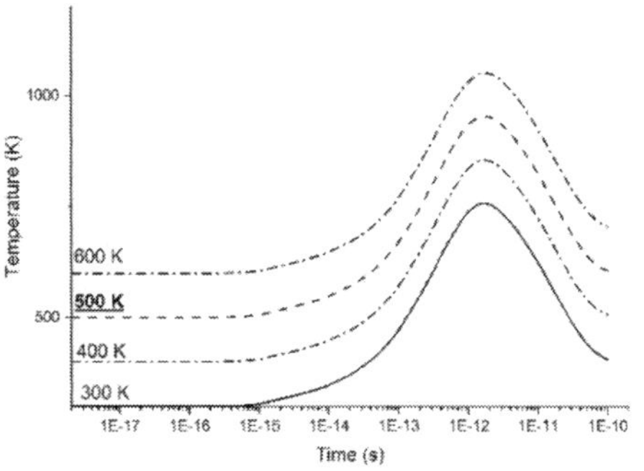

Fig. 5. Evolution of the lattice temperature of the Cu target as a function of time along the ion path for cases 2.

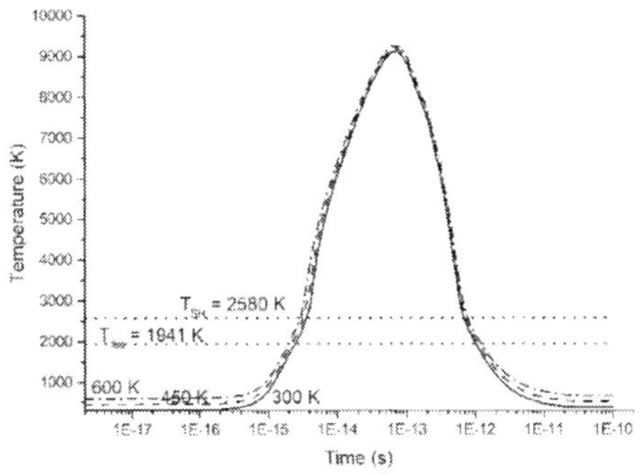

Fig. 6. Evolution of the la8.5ttice temperature of the Ti target as a function of time along the ion path for cases 2 and 3.

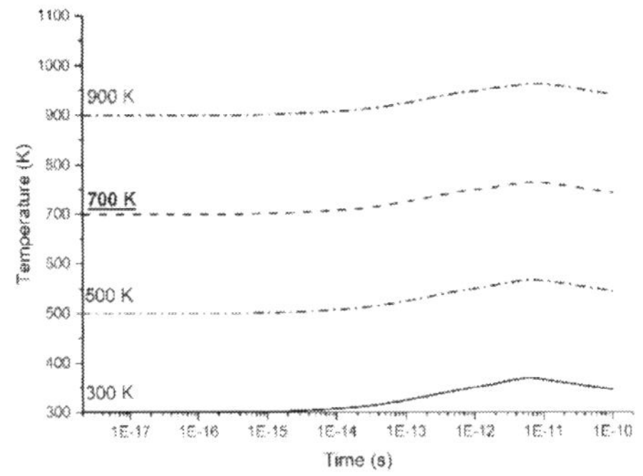

Fig. 7. Evolution of the lattice temperature of the Au target as a function of time along the ion path for cases 4.

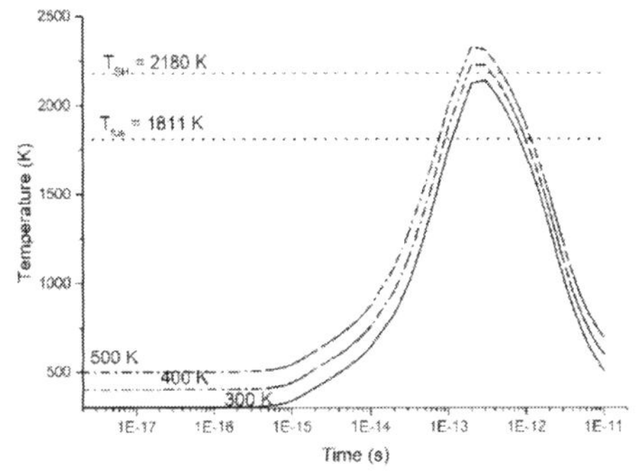

Fig. 8. Evolution of the lattice temperature of the Fe target as a function of time along the ion path for cases 5.

3 Discussion

The temperature distributions over time for the different systems show various features.

First the temperature is maximum at a different time according to the material, in the ranges [10^{-14}-10^{-13} s], [10^{-13}-10^{-12} s] or [10^{-12}-10^{-11} s] for Ti, "Fe, Au, Mg, Ni" or Cu respectively. This tendency is directly linked to the electron-phonon strength: the larger it is, the smaller are the heating and cooling times.

Secondly, the temperature rise varies according to either the material or the initial conditions. It ranges from ≈ 65 K for Au (Fig. 7) (or ≈ 300 K for Mg (Fig. 3), ≈ 450 K for Cu (Fig. 5)) to ≈ 8700 K for Ti (Fig. 6) and for Ni targets from ≈ 560 K in experiment 4 (Fig. 4 lower) to ≈ 1880 K in experiment 1 (Fig. 4 upper).

In experiment 4, the nickel and gold targets did not show any damage, and the model shows that the temperature increase due to spike did not reach the limit temperature (T_{SH}).

Compared to the first experiment, where the electronic stopping power of the Ni target was higher (7.8 keV/Å), its temperature increase could have influenced the Mg degrader temperature at equilibrium above the fusion temperature, explaining that it was slightly deformed as observed.

In the second experiment, the temperature distribution of the copper material (Fig. 5) is well below its limit temperature, by contrast with the titanium response to the spike where T_{SH} is reached fast, what could explain the rapidly degradation of the targets (Ti alone or Ti with Cu evaporated) at very low beam intensity.

As explained in [12] and regarding Fig. 9 for Ti targets, the latent track radius, defined as the radius of a cylinder of matter in which the energy necessary to melt is exceeded, is calculated to be about 8 nm. This value is in accordance with the one measured in [13] and calculated in [11] at dE/dx = 40 keV/nm. The radius of 8 nm corresponds to a fluence of $1.2*10^{12}$ particle per cm^2 inducing a first modification of the material, in other words before overlapping of spikes. This fluence is reached within 1 to 16 minutes with a beam of 0.1 pnA and a spot radius of 1 to 4 mm. Track overlapping can induce phase change in the titanium [14] which can affect drastically the target stability.

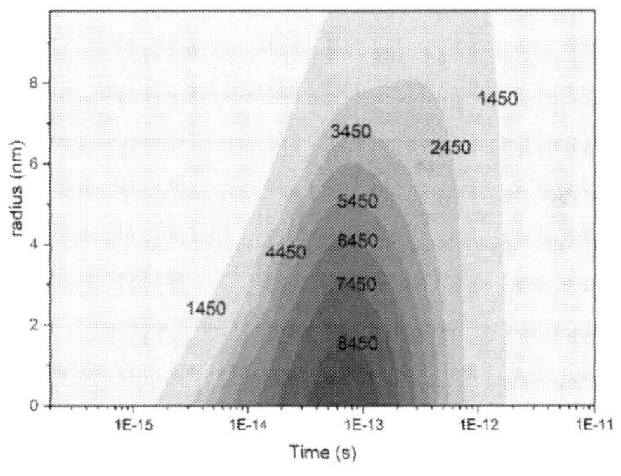

Fig. 9. Temperature distribution of the Ti target as calculated with the Thermal Spike Model.

The temperature rise is pronounced for Fe (≈ 1800 K, Fig. 8), and the higher value is near T_{SH} independent of the initial temperatures. However, the four iron targets used did not behave similarly under irradiation, one of them could sustain the beam up to an intensity of 80 nAe, the other ones broke at low intensity and within few hours. Apart from the potential "thermal spike" effect, one can speculate that other contributions to the damage of targets originates from the carbon/iron interface (carbon used as a backing) because of their different thermal expansion coefficients and/or the high sensitivity of iron to oxidation when it was on air before irradiation.

4 Conclusion

The thermal spike model, applied to some projectile-target combinations allows to understand that some metallic targets were deformed due to a fast and local temperature increase while others were insensitive. The model enables to predict reliably the sensitivity of targets to the electronic slowing down of heavy ion by considering fundamental parameters such as the energy loss and the electron-phonon coupling factor determined by some sputtering yields experimental studies. Preliminary estimations with this model prior to experiments will help to adjust some parameters, i.e. the beam energy and target material to prevent the damages.

References

1. B. A. Brown, Prog. Part. and Nucl. Phys. **47**, 517 (2001)
2. A. Dewald, O. Möller and P. Petkov, Progr. Part. and Nucl. Instr. Meth. A **67**, 786 (2012)
3. C. Dufour, A. Audouard, F. Beneu, J. Dural, J. P. Girard, A. Hairie, M. Levalois, E. Paumier, M. Toulemonde, J. Phys. Condens. Matter **5**, 4573 (1993).
4. M. Toulemonde, C. Dufour, A. Meftah, E. Paumier, Nucl. Instr. Meth. B **166-167**, 903 (2000)
5. Z. G. Wang, C. Dufour, E. Paumier, M. Toulemonde, J. Phys. Condens. Matter **6**, 6733 (1994)
6. C. Dufour, Z. G. Wang, E. Paumier, M. Toulemonde, Bull. Matter. Sci. **22**, 671 (1999)
7. C. Dufour, M. Toulemonde, *Ion Beam Modification of Solids* (Springer Ser. Surf. Sci. **61**, 63, 2016)
8. W. Assmann, M. Toulemonde, C. Trautmann, Topics Appl. Phys. **110**, 401 (2007)
9. M. Michel, note STP 570A, GANIL, Caen, 2016.
10. M. Michel, note STP 573A, GANIL, Caen, 2014.
11. M. Toulemonde, W. Assmann, C. Dufour, A. Meftah, C. Trautmann, Nucl. Instr. and Meth. B **277**, 28 (2012)
12. M. Toulemonde, C. Dufour, Z. Wang, E. Paumier, Nucl. Instr. and Meth. B **112**, 26 (1996)
13. H. Dammak, D. Lesueur, A. Dunlop, P. Legrand, J. Morillo, Radiat. Eff. Defects Solids **126**, 111 (1993)
14. M. Angiolini, H. Dammak, A. Dunlop, Philosophical Magazine Letters **82**, 81-89 (2002)

Multi-jet gas cooling of in-beam foils or specimens: CFD predictions of the convective heat-transfer coefficient

Gideon Steyn[1,*], *Christiaan* Vermeulen[2,**], and *Michael* Steyn[3]

[1]iThemba LABS, National Research Foundation, P.O. Box 722, Somerset West, 7129, South Africa
[2]Isotope Program, Los Alamos National Laboratory, Los Alamos, NM 87545, USA
[3]Stellenbosch University, Matieland, 7602, South Africa

Abstract. An experiment was designed to investigate the possible enhancement of the convective heat-transfer coefficient by utilizing multiple, parallel jets in the cooling of a small heated surface, such as typically induced by an accelerated ion beam on a thin foil or specimen. The hot spot was provided using a small electrically heated plate. It was found that heat-transfer calculations by means of simple empirical methods based on dimensional analysis are not useful in this case and that advanced computational fluid dynamics (CFD) mod-elling is essential to interpret the results. It is shown that enhanced convective cooling can indeed be obtained with a multi-jet configuration as compared to a single-jet configuration but only under very selective conditions. An improperly designed multi-jet configuration can also provide significantly reduced cooling relative to the single-jet case and the estimation of the behavior of any particular jet geometry is eminently non-intuitive. CFD provides acceptable quantitative results and seem to be the only tool available to gain an understanding of these complex flows where simple models and "rules of thumb" cannot be relied upon. – LA-UR-18-29455

1 Introduction

In a previous study (presented at INTDS 2016, see Ref. [1]), the focus was on the use of a single jet of gas in the convective cooling of thin in-beam materials, such as metal foils used as beam windows in a radionuclide production target station. Here we present a continuation of that work, where the investigation has been extended to multi-jet cooling of similar objects. This paper should therefore be read in conjunction with our previous work [1] to follow the full development of the topic. Of particular interest is the double-foil window concept, which is often employed to isolate the beamline vacuum from target materials. Such beam exit windows normally consist of two closely spaced foils through which a gas (e.g. helium) is circulated at a high flow rate to remove the heat induced by the beam.

What we are primarily interested in is the convective heat transfer from a small heated surface where an accelerated ion beam impinges on a target material. It is important to be clear on what is meant by "small heated surface" or "small hot spot" in the context of this study. There are many examples in the literature where an incident beam of ions generate a "small" heated area and one can easily familiarize with such situations, however, from a fluid dynamics perspective one should see the dimensions of a "small object" in relation to the dimensions of the inlet and exhaust ports that deliver and remove the coolant. When the heated area has similar or smaller dimensions relative to those of the coolant ports, it is appropriate to speak of a "small hot spot".

The idea of using multiple jets in the heat transfer from a small hot spot is in a certain sense an anomaly as multi-jet cooling had been introduced for the effective cooling of large heated surfaces. Typically, high-intensity heat transfer from a large surface area is achieved by means of an array of closely-spaced gas jets (usually air) impinging perpendicularly to the surface [2–4], such as in the drying of paper and textiles, tempering of glass, annealing of metal and plastic sheets, cooling of turbojet engine structures, etc. Examples in the literature are numerous and diverse. A common aspect is that these heated surfaces are typically very large in comparison to the jet diameter. Why, then, should one be interested in an investigation on multi-jet cooling of small heated surfaces?

The answer relates to a particular result reported in the literature by Chang et al. [5], where an enhanced convective heat-transfer coefficient was found with a 25-jet configuration (5 × 5 matrix) as compared to their single-jet measurements in a relatively small experimental assembly. These authors interpreted their result as due to an increased turbulence caused by the interaction of neighboring impinging jets, resulting in a higher average heat transfer, even though the local heat-transfer coefficients at the jet stagnation points were lower than in the single-jet case. Since the formalism by Chang and co-workers works well in the case of single-jet helium cooling of double-foil windows (our previous study – see Ref. [1]), we were curious about their multi-jet formalism and its possible application in convective cooling predictions for beam windows.

*e-mail: deon@tlabs.ac.za
**e-mail: etienne@lanl.gov

Many strategies are reported in the literature to enhance turbulence. So-called "passive turbulence promoters" are objects placed upstream from the nozzle exit, e.g. twisted tapes to cause swirl flow [6, 7], nozzle modifiers such as triangular tabs [8], etc. In the present study, the interaction of adjacent jets as a means to enhance turbulence has been investigated. We were primarily interested to see if the addition of a simple multiple-hole orifice plate could be used as a multi-jet nozzle in order to increase the turbulence, hence the overall heat-transfer coefficient.

At the 2016 INTDS Conference, we presented a paper on single-jet gas cooling of beam windows [1] in which we pointed out that certain empirical relations based on dimensional analysis have good predictive power. In addition, more advanced modelling based on computational fluid dynamics (CFD) proved useful to gain a better understanding of the turbulence and heat transfer inside such beam window assemblies. We also presented an experimental set-up designed to measure convective heat-transfer coefficients with a single gas jet. At that same time, we also had in our possession a set of measured data for multi-jet impingement heat transfer, which we did not present at INTDS 2016 because we did not then knew how to interpret those results. We are now in a better position to present those measurements as well as corresponding CFD simulations to assist with the interpretation.

In Sec. 2, the use of empirical relations based on dimensional analysis are briefly discussed for the case of multi-jet cooling. A description of the experimental set-up is presented in Sec. 3 as well as the new experimental data. Aspects of the CFD study are presented in Sec. 4, followed by some comments in conclusion in Sec. 5.

2 Dimensional analysis in multi-jet heat transfer – empirical formalisms

Many authors attempted to find appropriate empirical expressions for multi-jet heat transfer in terms of the Reynolds number, Nusselt number, Prandtl number, the geometric variables such as jet spacing, nozzle diameter, jet-to-surface spacing and the relevant fluid flow parameters. It transpired that success was rather limited and confined to very specific cases. There were simply too many correlations between too many variables, thus a general description proved to be elusive.

In our view, a seminal study was published in 1970 by Kercher and Tabakoff [2]. These authors employed the Buckingham pi theorem, which is the underlying the-orem in a whole range of heat-transfer expressions, such as the Dittus-Boelter, Sieder-Tate and Petukhov-Kirillov equations (see ref. [1] and references therein). Kercher and Tabakoff explored correlations of the following type in conjunction with measured heat-transfer data, employing a square array of round jets of the coolant (using forced convection of air in that particular study):

$$\mathrm{Nu}_D = \theta_1 \mathrm{Re}_D^m \mathrm{Pr}^n \left(\frac{X_n}{D}\right)^s \left(\frac{Z_n}{D}\right)^t \left[1 - \theta_2 \frac{G(X, I-1)}{G(H, I)}\right]^u, \quad (1)$$

where Nu_D is the Nusselt number evaluated with the jet diameter D as the characteristic dimension, Re_D is the Reynolds number evaluated with the jet diameter D as the characteristic dimension, Pr is the Prandtl number, X_n is the center-to-center nozzle spacing, and Z_n is the distance between the nozzle exits and the heated surface. The functions $G(X, I-1)$ and $G(H, I)$ in square brackets describe flow rates per unit area for total impingement and upstream crossflow, respectively, dependent on on yet further geometric/empirical parameters X, H and I which will not be elaborated in detail here as they are not required for the present discussion (but can be found in Ref. [2]). The quantities θ_1 and θ_2 are correlating constants (values to be determined by comparison with experimental data), while m, n, s, t and u are the exponents of the relevant dimensionless groups in the expression. Note that the first part of Eq. 1 is similar in form to Eq. 5 of our previous paper [1]. Usually the exponents assume constant values in this kind of study (similar to the correlating constants), at least as an approximation in particular regions of flow, pressure, temperature, or particular geometric constraints (normally expressed as ratios).

Amongst other problems with closed-form expressions such as Eq. 1, Kercher and Tabakoff showed that the group $(X_n/D)^s$ is not independent but a function of the Reynolds number. In particular, as a "slope parameter" the exponent s varies with Reynolds number. This results in the Buckingham pi theorem being violated. It was concluded that the heat transfer by their square array of impinging jets could not be correlated by power function expressions of dimensionless groups. Interestingly, various authors still attempted such correlations long after the Kercher and Tabakoff publication, probably because that work remained relatively unknown.

Since empirical methods based on dimensional analysis have little chance of success in providing a general description of heat transfer in multi-jet cooling configurations, the only remaining option to interpret measured data appears to be simulations based on CFD.

3 Measurements

3.1 Experimental set-up

The experimental set-up has been described in detail in our previous paper [1], therefore only some salient points are discussed here. A small circular heated surface (36 mm in diameter), a typical beam-spot size on our radionuclide production targets, was supplied by an electrically heated plate. Forced convection of the coolant (air) was supplied by a compressor. Diagrams of this assembly are shown in Figs. 1 and 2, details of which are described below.

A commercial electric element (manufactured by Swift Heat & Control: 230 V AC, 400 W) was embedded into an aluminium plate. The aluminium plate assembly was thermally insulated with mineral wool (supplied by Rockwool International, Denmark) except at the circular gas-cooled surface. The heated surface (circular hot spot) had a diameter of 36 mm. A k-type thermocouple was used to measure the temperature at the solid/gas interface. The power to the heater element could be adjusted with a Variac and Fluke instrumentation was used to read out the voltage and

https://doi.org/10.1051/epjconf/202022905002

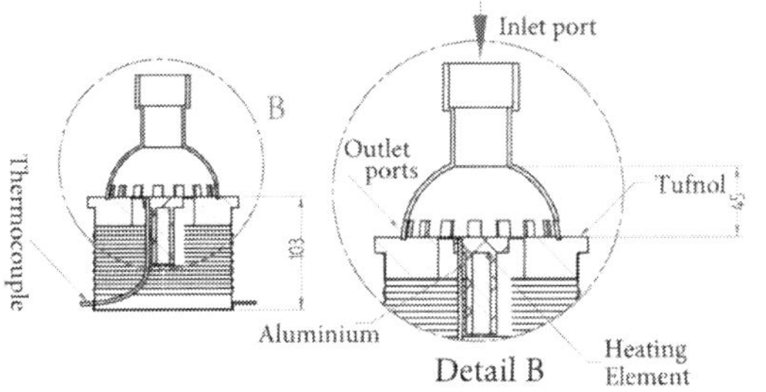

Fig. 1. Cross-sectional views of the electrically heated plate assembly for measuring convective heat-transfer coeffi-cients with multiple air jets. Note the single inlet port at the top and several rectangular outlet ports along the bottom edge of the hemispherical chamber.

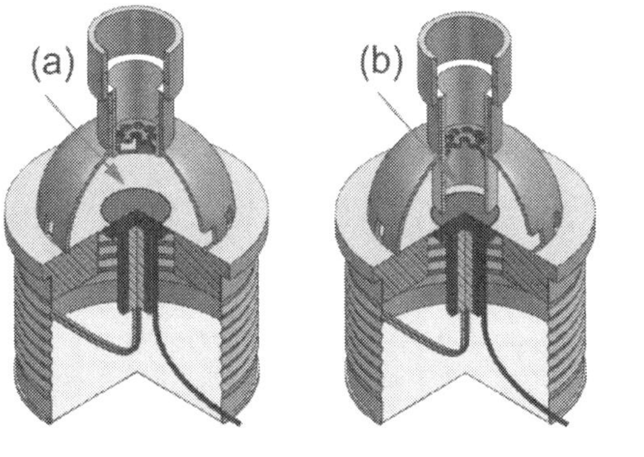

Fig. 2. (a) 3D view of the 90° hemispherical chamber (or dome) and electrically heated plate assembly with a cut-away to show the heated surface and multi-jet orifice plate. (b) The same view but with a flow guide installed (see text). The jet flow direction is perpendicular to the heated surface and spent gas is exhausted radially. The hot spot is shown in red.

current values. All temperature, voltage and current values could also be streamed to a computer in 1 second intervals. A small chamber with a hemispherical shape (or dome) fitted over the heated surface. The surface area adjacent to the heated surface was made of Tufnol, a material with more than two orders of magnitude lower thermal conductivity than aluminium. Also, the Tufnol plate was machined so that the material in contact with the aluminium plate was only 1 mm thick. This provided a single hot surface within a region predominantly at room temperature. Rectangular apertures along the bottom edge of the dome served as exit ports for the gas (eighteen exhaust apertures in total, each 10 mm wide and 11 mm high, located all around the circumference of the dome). An orifice plate with round holes could be fitted inside the inlet port at the top of the dome to provide the multiple jets (shown in green in Fig. 2).

Five such orifice plates were machined for the experiment, having 1, 3, 7, 10 and 16 holes, respectively. These plates were interchangeable inserts. Essentially, the experimental set-up was the same as in our previous study [1] except that an insert with holes could be fitted at the top of the dome. Each insert had a different number of holes but the total jet cross-sectional area remained the same. These inserts reduced the free flow area considerably compared to the previous single-jet case. The volume flow rate was therefore reduced for these measurements but the bulk linear velocity was nevertheless increased to 50 m/s (compared to 36 m/s in the case of the single-jet experiment). To be more precise, one of the inserts of the multi-jet experiment also had only one jet: Insert no. 1 (one jet, accurately machined to a diameter of 15.87 mm); Insert no. 2 (three jets, machined to a diameter of 9.175 mm each); Insert no. 3 (seven jets with a diameter of 6 mm each); Insert no. 4 (ten jets with a diameter of 5 mm each); Insert no. 5 (sixteen jets with a diameter of 4 mm each). In order to ensure that the total free flow areas of all the jet inserts were exactly the same, one of the holes had to be machined slightly larger or smaller than the others in a few cases. The idea behind the "flow guide" shown in Fig. 2 (b) will be explained later.

The thermal losses due to mechanisms other than forced convection were determined by covering the heated surface with insulating wool and measuring the electrical power versus temperature characteristics of the system for steady-state conditions without flow. These values were subtracted from the corresponding values obtained with flow. Heat-transfer rates were measured at surface temperatures of nominally 100, 150, 200, 250 and 300 °C.

3.2 Results and discussion

The measured convective heat-transfer coefficients are shown in Fig. 3 as a function of the number of jets. Also shown in Fig. 3 are calculated values according to the Chang, Dittus-Boelter, Sieder-Tate and Petukhov-Kirilov equations for a bulk linear velocity of 50 m/s. A wall temperature of 300 °C was adopted for the Sieder-Tate calculation. Similar to the approach in our previous study [1],

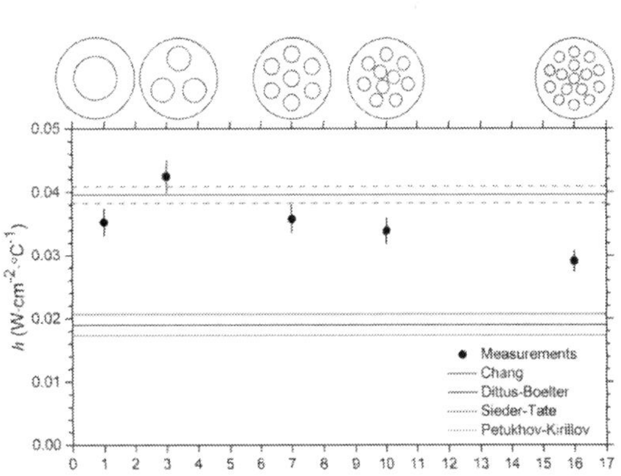

Fig. 3. Convective heat-transfer coefficient (h) plotted versus "number of jets" of multi-jet configuration. The bulk linear velocity of the air was nominally 50 m/s. Calculated values from other empirical formalisms are also shown, as indicated. The multi-jet inserts are shown schematically at the top of the figure.

the calculated values are shown as horizontal lines for purposes of comparison with the measured values, however, they are not to be understood as functions of the number of jets. The two red dashed lines are calculations using the Chang equations for flow rates adjusted by -6% and +6% relative to the nominal value of 35.4 m^3/h (50.0 m/s), respectively, as an indication of the measure of uncertainty. It is important to note that the single-jet Chang formalism was used for the result shown in Fig. 3. The 25-jet equation gives a larger heat-transfer coefficient, in contrast with the trend of these measurements. We concluded that the single-jet Chang formalism proved to be useful in our study but, in contrast, the multi-jet Chang formalism did not provide a useful approximation for the 25-jet case.

These results are most certainly very case specific. A significant enhancement in the convective heat-transfer coefficient relative to the single-jet value is found only in the 3-jet case. There is also a very slight enhancement with seven jets relative to one jet, however, this value is reduced considerably relative to the 3-jet value. This decreasing trend continues towards increasing numbers of jets. As before, the calculations with the Dittus-Boelter, Sieder-Tate and Petukhof-Kirilov equations underestimate the experimental values. The calculation with the single-jet formalism of Chang *et al.* is much closer to the measured single-jet value but the experimental value is nevertheless lower (Chang single-jet: h = 0.0396 W cm^{-2} $°C^{-1}$; measured single-jet: h = 0.0352 ± 0.0021 W cm^{-2} $°C^{-1}$). The measurement with the three-jet configuration was repeated several times and consistently resulted in higher values than obtained for the single-jet configuration (h = 0.0424 ± 0.0025 W cm^{-2} $°C^{-1}$).

It was speculated by Chang *et al.* [5] that the increased turbulence caused by the interaction between neigbouring jets may reduce the "jet potential core", leading to reduced stagnation point heat transfer but enhanced heat transfer away from the stagnation points. Based on this notion, there was some speculation at iThemba LABS that perhaps this kind of jet interaction may be having a "diverging" effect on the flow, leading to a significant spreading of the impinging coolant onto the uncooled surface adjacent to the heated surface. Such a spreading would then account for the observed reduction of the convective heat-transfer coefficient with increasing number of jets (for jet numbers larger than 3). In order to investigate this notion, a "flow guide" was installed, shown in Fig. 2 (b). This was just a short section of tubing below the jet insert, extending to 10 mm above the heated surface. Its purpose was to direct the bulk of the coolant onto the heated surface, preventing the primary jet from straying onto the adjacent uncooled surface. Measurements with the flow guide installed, however, produced nearly the same results as measurements without it. Thus, it was found that the flow guide had no significant effect at all.

One can indeed obtain an enhancement with a multi-jet configuration but, by the same token, one can also obtain a reduction. Consequently, the rule should be to *use multiple jets with caution!* As empirical formalisms have limited practical value in interpreting multi-jet heat-transfer data, we changed our focus to CFD modelling instead.

4 CFD modelling

4.1 Background and simulations

In our previous study [1] the code Autodesk CFD 2014 was used to simulate the experimental set-up for single-jet cooling. A later version – Autodesk CFD 2016 – was employed in the present work. The geometry of the set-up, as designed with the drafting code Solid Edge, was imported via AutoCAD Inventor into the CFD code. The coolant flow direction prior to impingement was perpendicular to the plane of the heated surface and a pressure of 1 bar at the exit ports was specified as a boundary condition.

For any CFD simulation involving turbulent flow, an appropriate "Turbulence Model" and "Advection Scheme" must be selected. Here advection refers to the transport of the coolant by bulk motion and, in a general sense, is the combination of convection and diffusion processes. Autodesk CFD 2016 has ten built-in turbulence models and five advection schemes. This gives a rather large number of possibilities, however, it is possible to narrow down the "sensible choices" considerably. From the online instruction manual and other sources, one can evaluate the "recommended uses" of the various turbulence models, e.g. some are recommended for external flows while others are intended for internal flows. One can eliminate turbulence models intended for natural convection, lower speed turbulent flows and buoyancy flows, etc. Five of the ten turbulence models are different variants of the Shear Stress Transport (SST) K-omega model. It was found that the standard SST K-omega model was sufficient for our needs, therefore the special versions (namely SAS [Scale Adaptive Simulation]; RC [Rotation and Curvature correction] Smirnov-Menter; RC Hellsten; and DES [Detached Eddy Simulation]) were not required. Together with the K-epsilon model (the default model in the code), it was possible to narrow the turbulence models down to two. The same applies to the advection schemes. Three of them are appropriate for our kind of analysis, namely the Monotone Streamline Upwind scheme, Petrov-Galerkin scheme and Modified Petrov-Galerkin scheme (code named ADV 1, ADV 2 and ADV 5, respectively). The Flux Based Scheme (ADV 3) is specially tuned for external flow problems involving significant drag, thus not recommended for the internal flow situation of the present study. Similarly, ADV 4 is specially tuned for flows in long, narrow ducts, thus could be excluded from our investigation. ADV 1 is the default scheme and considered the "work horse" scheme by the developers of the code, recommended as the starting point for most analysis types. ADV 5 is reported to give better accuracy when there are splitting and secondary flows of the coolant, thus one may expect it to be useful in a multi-jet application. It is also reported to be more stable than ADV 2, therefore we chose the modified instead of the standard Petrov-Gelerkin scheme. (Several excellent YouTube videos available on advection schemes and other aspects of the code.)

One of the great strengths of Autodesk CFD is that it is an integrated development environment, which simplifies the work flow considerably. This avoids, for example, mesh generation with a separate pre-processor and displaying of the results with a separate post-processor, as is typically the case with most CFD software. In our simulations, Automatic Mesh Sizing was selected to generate the initial mesh. This mesh can be made finer or coarser with a Size Adjustment Slider. The mesh can be viewed in three dimensions by performing rotations, translations, and zooming in and out. Prior to the start of any simulation run, the code performs several checks on the mesh and reports potential problems and/or errors. Other powerful meshing features include Boundary Mesh Enhancement and Enhancement Blending, both of which will be explained in more detail below.

In our previous study on single-jet gas cooling [1], the SST K-omega model was found to give results in better agreement with experimental data than the default K-epsilon model. In the present study, convergence issues/errors were sometimes experienced with the K-epsilon model, especially with increasing numbers of jets. These problems were not found with the SST K-omega model. Consequently, we decided quite early in the investigation to discontinue using the K-epsilon model. Finally, only the SST K-omega model in conjunction with the ADV 1 an ADV 5 advection schemes were retained for our analysis, reducing the 50 possible combinations of "turbulence model + advection scheme" to only two.

In the SST K-omega model, no "damping functions" or "wall functions" are used to describe the strong attenuation of the normal component of turbulence near the wall boundary, as is done in the K-epsilon model. Instead, the turbulence is simulated all the way to the wall, requiring the mesh to be very fine in the boundary layer region. The default number of mesh layers at a wall boundary is 3 but it is recommended to increase this number when using the SST K-omega model. This is done by using the Boundary Mesh Enhancement feature of the code. The 2014 version of the code allowed the number of boundary mesh layers to be increased to 10, while the 2016 version allows up to 15 layers. The price to pay for increasing the number of layers is a large increase in the total number of elements and the simulation becoming much more resource intensive, requiring a significantly longer execution time. In order to keep the total number of elements below 3×10^6 in the case of 16 jets, the number of mesh layers at the wall was increased from 3 to only 8. While 10 layers is normally recommended for the SST K-omega model, it was found that there was not much difference in the results by adopting a parameter value of 8. With a top-end core-i7 PC, this choice would result in the completion of a run in under 12 hours, therefore a second run could be started on the same day and completed during the night. The number of elements and the execution time increased with increasing number of jets. The total number of elements varied between about 800 000 for the single-jet case to 3×10^6 for the 16-jet case.

Enhancement Blending was selected in all our simulations. The code default is with Enhancement Blending disabled, which results in an instantaneous transition from a highly anisotropic wall region to an isotropic interior region. A more gradual transition in element size between wall regions and interior regions is obtained when Enhancement Blending is enabled, which is often recommended for simulations with the SST K-omega models.

The boundary conditions were as follows: Inlet air temperature - 20°C; Outlet air pressure - 1 bar; Bulk flow rate - 35.4 m^3/h. The default convergence criterion is halfway between "loose" and "tight" in the code, with a slider to select different values. The notes in the code recommend tightening the convergence criteria once a new problem runs well. The slider was typically adjusted from the 50% position to the 80% position, i.e. towards a tighter convergence criterion. This was found to be a good compromise. With the tightest convergence criterion (slider at 100%) some simulations took significantly longer to complete. With CFD simulations, refinements usually have a price to pay in terms of extended execution time.

4.2 Results and discussion

The measured convective heat-transfer coefficients are compared with the corresponding CFD results in Fig. 4. In the cases of single-jet and 3-jet cooling, the results obtained with the Monotone Streamline Upwind scheme (i.e. the ADV 1 advection scheme) are in very good agreement with the measurements but towards an increasing number of jets one observes an underestimation. In contrast, the Modified Petrov-Galerkin scheme (i.e. the ADV 5 advection scheme) produces an overestimation for the single-jet and 3-jet heat transfer but good agreement with the experimental values for the 7-jet, 10-jet and 16-jet cases. In Fig. 4, the closest CFD results to the measurements have been filled with magenta colour. Indeed, the magenta symbols show good agreement with the experimental values. This is very interesting because, from the perspective of the fluid flow, an increase in the number of jets leads to

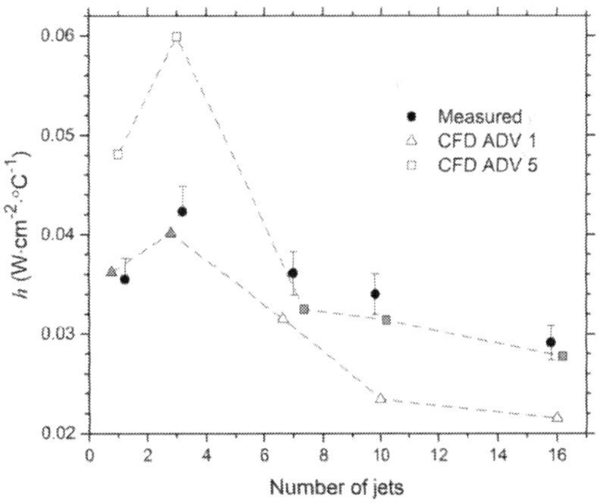

Fig. 4. Convective heat-transfer coefficient versus number of jets. The measured values are compared with CFD results obtained using two different advection schemes, as indicated (see text). Note that some results are slightly displaced on the x-axis to prevent overlap of the symbols. The CFD results closest to the measured values are indicted with a magenta fill. The dashed lines are to guide the eye.

more complexity in the flow in terms of separation and re-combination. The ADV 5 advection scheme is reported to be superior in such cases. In the simpler cases of single-jet and 3-jet flow, however, the ADV 1 advection scheme gives better agreement. In the case of 7 jets, ADV 1 and ADV 5 give very similar results. Whether this is truly significant or simply a coincidence is debatable, however, it is not unexpected that ADV 5 should give better results in cases where separation and recombination in the coolant become more abundant. It is not clear to us whether one should expect ADV 1 to be superior in the cases of simpler flows, however, in this particular study such a result was indeed obtained. Both sets of CFD simulations exhibit a maximum heat-transfer coefficient with 3 jets, in agreement with the experimental results.

The above results were obtained with thin (3 mm thick) orifice plates to form the jets. In many applications, however, jet nozzles are rather elongated in shape. We asked ourselves whether jet length would play a significant role. We therefore performed a set of CFD simulations with elongated jets. The nozzle length was increased to 100 mm (i.e. many jet diameters in length for all cases), thus one would expect the flow to become fully developed in a jet prior to reaching the nozzle exit. These thick nozzle inserts extended upstream into the tube through which the coolant entered the dome. This is evident in Fig. 5, which shows a shaded wire view of the fluid mesh for the 7-jet case (note that all solid structures were made invisible in this view in order to show only the fluid.) Figure 6 compares the CFD results obtained with thin jet inserts and thick jet inserts. It is evident that the nozzle length has only a minor effect on the results. Note that ADV 5 was used in these simulations. It is interesting that slightly higher heat-transfer coefficients have been obtained with the thinner nozzles but the reason for this is not clear.

Next, we questioned the effect of the distance of the nozzle exits to the heated surface. CFD simulations were therefore also performed at two smaller nozzle-to-heated-plate distances to investigate this. The three sets are compared in Fig. 7. Note that advection scheme ADV 5 was again selected for this comparison. It is clear that the nozzle-to-heated-plate spacing has a significant effect on the results. It seems generally advantageous to place the nozzles closer to the heated surface. At the smallest nozzle-to-wall separation (10 mm) the heat-transfer coefficient was found to be larger in the 7-jet case. Thus, the maximum shifted from the 3-jet to the 7-jet configuration by placing the jets closer to the surface.

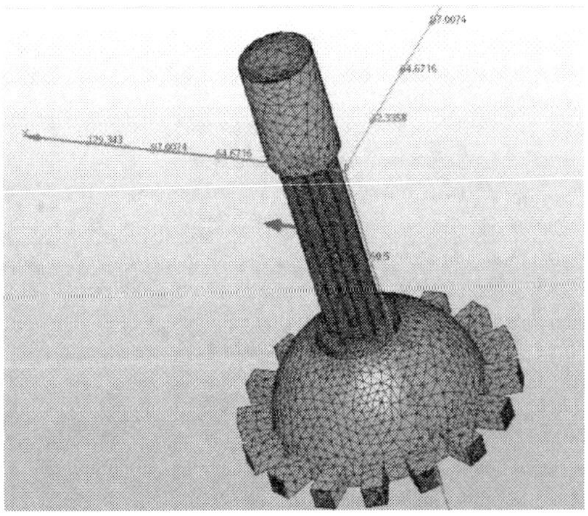

Fig. 5. Shaded wire view of the fluid mesh in the 7-jet assembly with elongated jets.

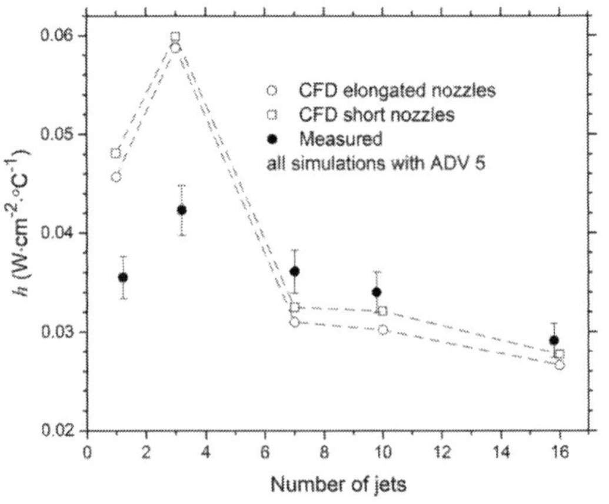

Fig. 6. Convective heat-transfer coefficient versus number of jets. The CFD results with elongated nozzles (100 mm) are com-pared with CFD results obtained with a thin (3 mm) nozzle insert. The dashed lines are to guide the eye. The measured values (ob-tained with thin nozzles) are also shown for comparison.

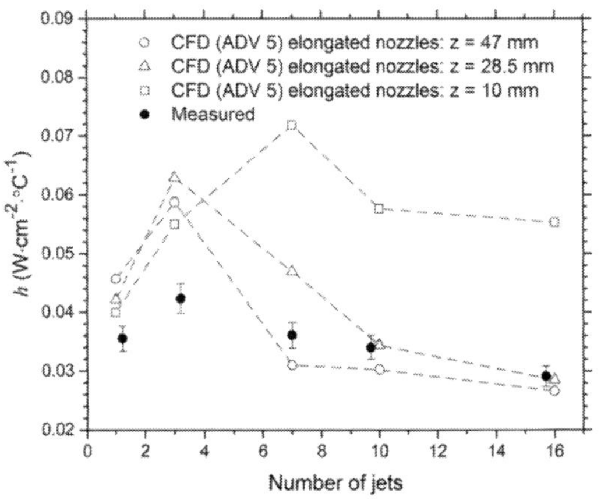

Fig. 7. Convective heat-transfer coefficient versus number of jets. The CFD results with elongated nozzles (100 mm) are compared at three nozzle-to-heated-surface distances, as indicated. The dashed lines are to guide the eye.

In the final set of simulations performed, the flow rate was increased from 35.4 m^3/h to 100 m^3/h. The results are shown in Fig. 8. As expected, higher heat-transfer coefficients have been obtained with the higher flow rate, however, the trend of the heat-transfer coefficient with respect to number of jets remained the same. The difference between the two sets of results seems to be a simple scaling factor, to a good approximation.

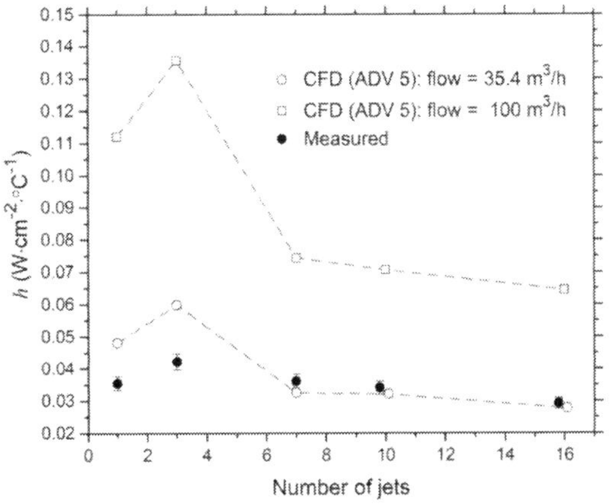

Fig. 8. Convective heat-transfer coefficient versus number of jets. The CFD results with elongated nozzles (100 mm in length) are compared at a nozzle-to-heated-surface distance of z = 47 mm but for two different flow rates, as indicated.

5 Conclusion

It has been shown that the use of multiple jets can enhance turbulence, hence the convective heat-transfer coefficient. Whereas Chang and co-workers found an enhancement with a 5 × 5 square array with respect to their single-jet case, in our experiment we found an enhancement with respect to the single-jet case only with our 3-jet and 7-jet configurations, whereafter the further increase of the number of jets lead to a monotonically decreasing heat-transfer coefficient, and a worsening relative to the single-jet case. Geometry clearly plays an important role, making it very difficult to use dimensional analysis to relate the heat transfer to the various hydrodynamic properties. In this regard, the cross flow of spent gas towards the exhaust ports play a major role. Due to these complexities, CFD simulations in conjunction with experiments seem to be the only alternative to study the behavior of such systems.

In this study, which should be seen as a continuation of our previous study reported at INTDS 2016, we endeavoured to utilize multi-jets as a way to improve the cooling of window foils on targets and beamlines with gas as the coolant. Since the flow i n s ubmerged m ulti-jet systems is complex and non-intuitive, one can easily get it wrong. Turbulence enhancement is indeed possible but one can also achieve the opposite, i.e. a reduction in turbulence in the region where most of the heat transfer takes place.

The SST K-omega model, within the context of CFD simulations, shows good potential as a means to study multi-jet cooling of small heated objects. It provided convective heat-transfer coefficients in good agreement with the experimental results of this study but there is a clear dependence on the choice of advection scheme. The Modified Petrov-Galerkin scheme provided good agreement in cases where the number of jets is higher (more than 3). In the simpler single-jet and 3-jet configurations, very good agreement was obtained with the Monotone Streamline Upwind scheme. The latter advection scheme seems to be better for simple flow geometries while the former one better for more complex, separated flows.

Multi-jet cooling in targetry clearly has potential but must be implemented with care. There are no "rules of thumb" or readily available approximations. A combination of experimental investigation and CFD modelling is required to gain a better understanding of the heat transfer.

Lastly, we need to point out that our measurements were performed using air as coolant. This was done for the sake of ease and also because the assembly was not helium tight. Normally, helium would be used as it is superior to air as a coolant gas, in addition to not forming problematic activation products.

This work was financially supported by the National Research Foundation (NRF Grant no. 85507).

References

[1] G.F. Steyn, C. Vermeulen, *INTDS2016, AIP Conf. Proc.* **1962** (2018) 030020-1 – 030020-11
[2] D.M. Kercher, W. Tabakoff, *J. Eng. Power* **92** (1970) 73–82
[3] H. Martin, *Adv. Heat Transfer* **13** (1977) 1–60
[4] Y. Shan, J. Zhang and G. Xie, *Int. J. Heat Mass Transfer* **86** (2015) 832–842
[5] C.T. Chang, G. Kojasoy, F. Landis and S. Downing, *Int. J. Heat Mass Transfer* **38** (1995) 833–842
[6] H.Q. Yang, T. Kim, T.J. Lu and K. Ichimiya, *Int. J. Heat Mass Transfer* **53** (2010) 4092–4100
[7] K. Nanan, K. Wongcharee, C. Nuntadusit and S. Eiamsa-ard, *Int. Com. Heat Mass Transfer* **39** (2012) 844–852
[8] Y. Yu, J. Zhang and H. Xu, *Int. J. Heat Mass Transfer,* **72** (2014) 222–233

Rhenium and iridium targets prepared using a novel graphene loading technique

John P. Greene[1,*], *Matthew* Gott[1], *Richard L.* Fink[2], and *Igor* Pavlovsky[2]

[1]Physics Division, Argonne National Laboratory, Lemont, Illinois, United States
[2]Applied Nanotech, Inc., Austin, Texas, United States

Abstract. For accelerator targets, graphene films are an excellent material choice due to their high thermal conductivity, high temperature tolerance, low outgassing, mechanical integrity, and ease of handling. A variety of targets have been produced using graphene material as a backing or a host matrix. One of the unique advantages of the graphene film fabrication process is the capability to embed target materials, including refractory metals, in the nanoparticle form into a host graphene matrix during target preparation. Targets of ^{nat}Ir and ^{nat}Re have been fabricated as nanoparticle loaded graphene targets for use in nuclear physics research. We have obtained beam time to evaluate target performance as well as production yields and nuclear decay properties via the ^{nat}Re(a,2n)^{186}Ir and ^{nat}Ir(a,3n)^{194}Au reactions, respectively. These rhenium and iridium targets will be irradiated using the ATLAS accelerator and gamma rays measured in-place using the high-precision gamma-ray spectroscopy capabilities of Gammasphere and further analyzed using a multi-parameter detector system. Future plans include the preparation of isotopic targets of these two elements.

1 Introduction and motivation

With cancer incidence rates continuing to climb, research into advanced, diagnostic and therapeutic radiopharmaceuticals is critical to meeting future needs [1]. In this regard, a theranostic drug allows for low dose imaging and dosimetry estimates followed by a tailored course of therapeutic treatments based on the diagnostic evaluation. The success of this approach is dependent on high specific activity radionuclides. Some recent studies into iridium-based anti-cancer drugs have used photosensitizing complexes to attack cancer cells [2-4]. Incorporation of the Auger-emitting ^{186}Ir ($t_{1/2}$ = 16.64 h) could lead to a synergistic killing effect improving the overall treatment by reducing the required dose of the chemotherapeutic agent. Additionally, its gamma emissions (γ = 296.9 keV, 62.3%; 137.2 keV, 41.4%) are suitable for SPECT imaging and could have an immense impact on determining dose and clearance characteristics. Cytotoxic gold compounds have also been evaluated as anti-cancer drugs [5]. Auger-emitting ^{194}Au ($t_{1/2}$ = 38.02 h) can provide the same synergistic killing effect and SPECT imaging (γ = 328.5 keV, 60.4%; 293.5 keV, 10.6%) for gold-based anti-cancer drugs.

The experiment described here would attempt to produce high specific activity radio-iridium and -gold via the ^{185}Re(a,2n)^{186}Ir [1087 mb; 35 MeV] and ^{193}Ir(a,3n)^{194}Au [1101 mb; 35 MeV] reactions, respectively. Though we will utilize natural abundance targets in these initial experiments, these isotopes are highly naturally abundant (37.40% for ^{185}Re and 62.7% for ^{193}Ir). To date, experimental data in the EXFOR database is extremely limited for these reactions, particularly at the alpha energy of interest. Most the data relies on measurements performed during the 1990s [6-9]. Theoretical production yields were calculated based on a 1 mg/cm^2 target and a beam current of 1 pnA. Production rates of 12.9 µCi/nAh and 20.8 µCi/nAh are expected for ^{186}Ir and ^{194}Au, respectively.

2 Description of the graphene loading technique

Rhenium was first discovered in platinum ores by Noddack, Tacke and Beer in 1925 [10]. The origin of the name comes from the Greek word "Rhenus" meaning river "Rhine." Rhenium does not occur free in nature or as a compound or mineral though it is widely spread throughout the earth's crust at a concentration of 0.001 ppm. The element is silvery-white with a metallic luster and when annealed, rhenium is very ductile and can be bent, coiled or rolled. Its atomic weight is 186.2, density of 21.02 g/cm^3 and a melting point of 3180° C. The isotopes are ^{185}Re at 37.40% and ^{187}Re at 62.60% natural abundance, respectively and available as highly enriched metal powder.

Iridium was discovered in 1803 by Tennant [10] in residue left from platinum dissolved in aqua regia and the name is derived from its salts which are highly colored. It occurs uncombined in nature and is found with platinum in alluvial deposits. The element is white, similar to Pt but with a slight yellowish cast and is hard

* Corresponding author: greene@anl.gov

and brittle making it difficult to machine, form or work. Iridium is the most corrosion resistant metal known. Its physical properties include; atomic weight 192.22, a melting point of 2410° C and a density of 22.42 g/cm^3. There are two naturally occurring isotopes; ^{191}Ir with 37.3% natural abundance and ^{193}Ir with 62.7% natural abundance. They are also available as highly enriched metal powder.

Maier-Komor first reported on the preparation of Re targets on thin carbon backings by the method of electron bombardment [11] and similarly by Demaret onto aluminum backings [12]. Rhenium metal is also easily rolled. Isotopically enriched iridium targets have been prepared by McDaniel (*et al.*) [13]. Richaud produced targets via a centrifugal method [14] and Kobisk (*et al.*) suggested preparing, for example, ruthenium and iridium targets by embedding in a Cermet matrix [15]. Iridium, like osmium is too brittle to be rolled or formed.

The refractory metal targets are notoriously difficult to produce due to their extremely high melting points. Applied Nanotech, through an SBIR (Small Business Innovation Research), has developed a novel method to produce thin refractory metal targets imbedded in a graphene matrix and contain 1:1 metal: carbon by weight [16]. As these with rhenium and iridium loaded graphene targets have never been exposed to heavy-ion beam irradiation, testing their in-beam stability and performance was necessary. The targets were therefore first bombarded with 0.5 pnA α beam to evaluate target integrity and to produce activity for gamma spectroscopy before enabling higher power irradiation.

One of the unique advantages of this new process is the capability to embed target materials, including the refractory metals, in the nanoparticle form into a graphene film during the film fabrication. Iridium nanoparticle dispersions were prepared in ethylene glycol by reduction of hexachloroiridate ($H_2IrCl_6{*}H_2O$) at elevated temperatures up to 150° C. The average particle obtained was in the range of 170 nm to 600 nm with the particle size distribution measured using a Nanotrac particle analyzer. Also, a sample of Ir nanoparticles was purchased from American Elements having nanoparticles with average size of <100 nm. The material was dispersed in deionized water water to obtain 10 mg/ml Ir stock dispersion.

To prepare Ir-loaded targets, the ^{nat}Ir nanoparticle dispersions were added and mixed with the prepared graphene solutions before the final filtration. A range of different thicknesses and loadings of Ir in the graphene matrix were obtained. To ensure that the Ir-loaded graphene targets keep good mechanical integrity in beam, the mass loading of Ir of 1:1 over graphene was not exceeded (higher Ir loadings might interfere with van der Waals attraction forces between graphene crystallites). Area densities of the obtained films were in the range of 0.4 to 0.6 mg/cm^2.

Several 22 mm diameter ^{nat}Ir targets were made from dispersions using a 125 mm syringe pump-based pressure filtration system (Fig. 1). The filtration systems and polymer filters were purchased from Cole Parmer. The targets have Ir loading of 1:1 with respect to the mass of graphene, where the areal mass density of graphene was in the range of 0.2 to 0.3 mg/cm^2. Quadratic samples of 40x40 mm^2 were then cut from the 125 mm disk samples to prepare targets.

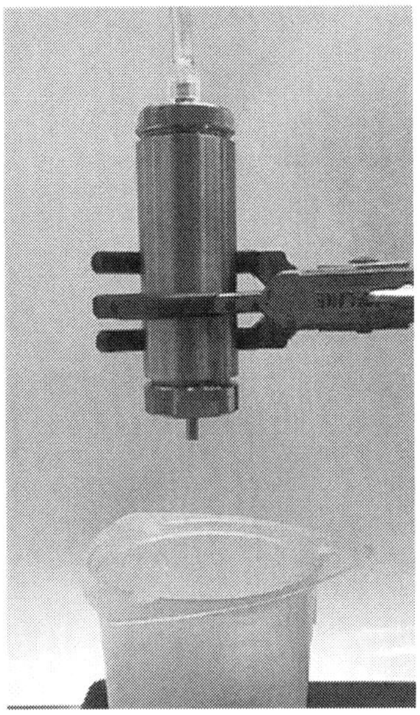

Fig. 1 Stainless steel, one inch diameter pressure filtration system

This process was repeated to produce graphene targets loaded with ^{nat}Re nanoparticles. To prepare Re nanoparticles, ammonium perrhenate (APR) powder, purchased from Sigma Aldrich, was reduced in a tube furnace. Approximately 0.5 g of this APR powder was spread uniformly over 10 cm^2 of a tungsten boat. The boat was then moved into the center of the tube furnace, and 100 sccm N_2 flow was used to purge the tube for 20 minutes. The oven temperature was then ramped at 10 °C/m to 320 °C, and held at this temperature for 1 hour. After dark gray ReO_2 was formed in the boat, 50 sccm H_2 was introduced to further reduce the rhenium oxide to metal powder at the same temperature for another 2 hours. The aqueous powder dispersion was then processed in a SpeedMixer and sonicated. Natural rhenium targets were produced the same way as Ir targets, using a 1 inch diameter stainless steel pressure filtration system. Table 1 lists the Re and Ir targets that were prepared on target frames and used for beam irradiation (Fig. 2). Future plans include producing these graphene targets using enriched isotopes available from Oak Ridge National Laboratory.

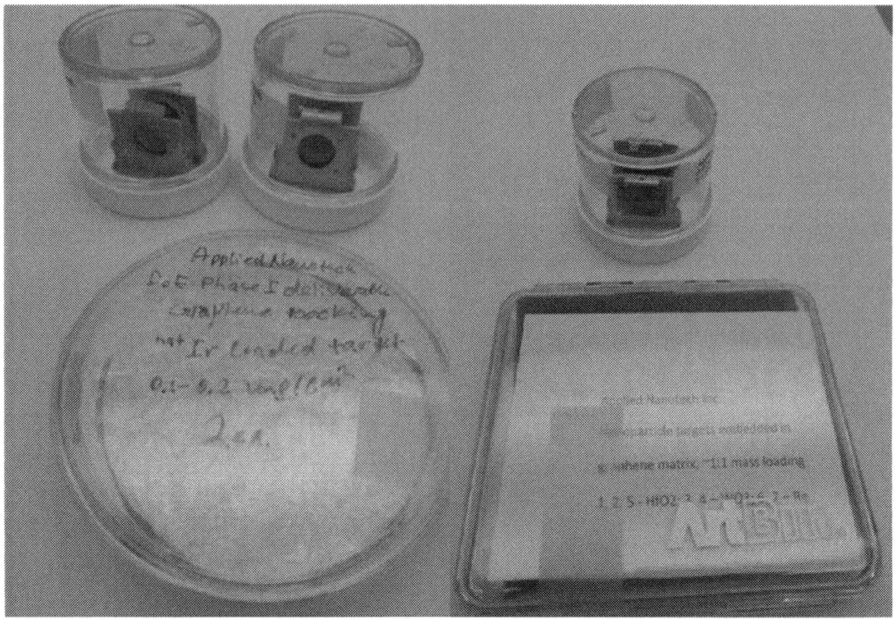

Fig. 2 - ^{nat}Re mounted targets (left) and ^{nat}Ir mounted targets from foils supplied by Applied Nanotech, Inc.

Table 1. Prepared Targets and Thicknesses

TARGET	TARGET THICKNESS (mg/cm^2)	NUMBER OF TARGETS
^{nat}Re	0.6 (0.3 Re + 0.3 C)	2
^{nat}Ir (American Elements)	0.4 (0.2 Ir + 0.2 C)	4
^{nat}Ir (ANI)	0.6 (0.3 Ir + 0.3 C)	4

3 Experimental results and discussion

Two days of requested beam time were used for evaluation of ^{186}Ir and ^{194}Au excitation functions and production yields via the ^{nat}Re(a, 2n)^{186}Ir and ^{nat}Ir(a, 3n)^{194}Au reactions, respectively. Irradiation of these novel rhenium- and iridium-loaded graphene targets was accomplished using the ATLAS superconducting accelerator by two methods. First, with single targets in-beam experiments providing measurement of the decay gamma rays using the high-precision gamma-ray spectroscopy capabilities of Gammasphere. Next, a set of targets were arranged in a "stack" using Cu foils interlaced between the Ir and Re targets to degrade the beam energy – providing also a known flux determination for later cross-sectional data extraction. For this, several Cu foil targets with thicknesses of 2.2 mg/cm^2 and 4 mg/cm^2 were first prepared by mechanical rolling [17]. This target stack was bombarded for 5 hours at 5 pnA of alpha beam with energies between 32-38 MeV, degrading as it traverses the target stack. After beam bombardment, the target stack was disassembled, enabling the measurement of an excitation function of this reaction using several targets of each element. Examination of the targets after irradiation showed no visual damage (Fig. 3).

The ^{194}Au and ^{186}Ir reaction products themselves will be stopped inside the targets, which can be used for in-beam and off-line activity measurements. Using in-beam detection of prompt gammas, we can determine the production rates for the stable isotopes in addition to those of the radioisotopes. This is critical for determining the specific activity and purity of the produced medical isotopes and is useful in testing model predictions. In-place determinations of the target activities at EOB was done using Gammasphere. In order to quantify the absolute gamma efficiency for these measurements, a calibrated standard source was located next to the target within the experiment chamber. The gamma rays from the reaction products and from this known source were then measured at the same time, providing high-precision gamma spectroscopy data. The gamma intensities from each Gammasphere run can be used to precisely calculate the production yields and determine the reaction cross-sections at the degraded beam energy. Some data was also collected with both Cu and Ir targets at the same time to determine if this method can be used for similar measurements with other reactions in the future. These offline decay measurements enabled validation of production rates and provide a baseline count for further irradiations.

Fig. 3 ^{nat}Ir-loaded graphene target, 0.6 mg/cm^2, after beam irradiation

4 Conclusion

One of the unique advantages of the graphene film fabrication process is the capability to embed target materials, including refractory metals, in the nanoparticle phase into a host graphene matrix during target preparation. ^{nat}Ir, ^{nat}Re, as described above, and $^{nat}WO_3$, and $^{nat}HfO_2$ nanoparticle-loaded graphene targets were fabricated for use in nuclear physics research. Thin graphene/refractory metal targets with thicknesses of 0.4 - 0.6 mg/cm^2 were produced. The targets were easy to handle and to attach to Gammasphere target frames. Targets of ^{nat}Re and ^{nat}Ir were successfully irradiated without being damaged and the gamma-rays measured offline were used to determine production cross sections for the radionuclides of interest. The resulting cross sections, impurity production rates, and potential thick isotopic target production yields will be forthcoming.

This material is based upon work supported by the U.S. Department of Energy, Office of Science, Office of Nuclear Physics, under Award Number DE-SC-0017208 and Contract No. DE-AC02-06CH11357. This research used resources of ANL's ATLAS facility, which is a DOE Office of Science User Facility.

References

1. Meeting Isotope Needs and Capturing Opportunities for the Future: The 2015 Long Range Plan for the DOE-NP Isotope Program, NSAC Isotopes Subcommittee, July 2015
2. P Zhang, CKC Chiu, H Huang, et al., Angew. Chem. Int. Ed., **56**, 14898-14902 (2017)
3. N Wu, J-J Cao, X-W Wu, et al., Dalton Trans. **46**, 13482-13491 (2017)
4. H Xiang, H Chen, HP Tham, et al., ACS Appl. Mater. Interfaces **9**, 27553-27562 (2017)
5. L Messori, F Scaletti, L Massai, et al., Chem. Commun., **49**, 10100-10102 (2013)
6. M Ismail, Pramana J. Phys., **50**(2), 173-189, (1998)
7. JD Stickler, KJ Hofstetter, Phys. Rev. C, **9** (3), 1064-1071 (1974)
8. MK Bhardwaj, IA Rizvi, AK Chaubey, Phys. Rev. C, **45** (5), 2338-2342 (1992)
9. T Szuecs, GG Kiss, Gy Gyuerky, et al., Phys. Rev. C **776**, 396-401 (2018)
10. RC Weast (1980) Handbook of Chemistry and Physics, 56th ed. CRC Press, Cleveland
11. P Maier-Komor, Proceedings of the 1974 Annual Conference of the INTDS, Chalk River, Ontario, 70 (1974)
12. Demaret, INTDS Newsletter Vol. **20** No. 1, 7 (1993)
13. EW McDaniel, Proceedings of the 1974 Annual Conference of the INTDS, Chalk River, Ontario, 151 (1974)
14. JP Richaud, Nucl. Instr. and Meth. **167** ,91 (1979)
15. EH Kobisk, TC Quinby and WS Aaron, Preparation of Nuclear targets for Particle Accelerators, J Jaklovsky, Ed. Plenum Press, New York, 127 (1981)
16. *To be published*
17. FJ Karasek, Nucl. Instr. and Meth. **102** , 457 (1972)

Second generation degrader foil for the CARIBU project

Matthew Gott*, *John* Greene, *Brad* DiGiovine, *Guy* Savard, and *Bruce* Zabransky

Physics Division, Argonne National Laboratory, Lemont, Illinois, United States

Abstract. The Californium Rare Ion Breeder Upgrade (CARIBU) project utilizes ^{252}Cf to provide beams of neutron-rich nuclides with intensities not currently available at other facilities. The electroplated ^{252}Cf source is positioned inside a large helium gas catcher, where the ejected fission fragments lose their energy and are slowed. Prior to entering this gas catcher, the ions first pass through a gold cover foil to contain self-sputtering recoil emissions and then through an aluminium degrader foil, where much of their residual energy is reduced. In the past due to production limitations, a less than ideal cylindrical shaped degrader was utilized. This resulted in non-uniform energy loss as the ions passed through the degrader. With the advent of 3D printing, a new hemispherical degrader was prepared to enable a more uniform energy loss. The design, production, and assembly will be discussed.

1 Introduction and motivation

Beams of exotic neutron-rich nuclei are of high interest for nuclear physics research studying phenomena such as nucleosynthesis of heavy nuclei and nuclear structure of nuclides approaching the neutron dripline. The Californium Rare Ion Breeder Upgrade (CARIBU) project utilizes ^{252}Cf to provide radioactive beams not produced in the low-energy fission of uranium as well as producing elements that are difficult to extract using standard ISOL techniques [1]. CARIBU provides beams of neutron-rich species to the Argonne Tandem Linear Accelerator System (ATLAS) where they can be accelerated up to ~ 10 MeV/*u*. The ^{252}Cf source is positioned inside a large RF gas catcher filled with high-purity helium, where the fission fragments lose energy and are slowed down. Once extracted, the ions are accelerated to the isobar separator, where two 60° magnets are used to isolate the species of interest; this high-purity beam is then sent directly or reaccelerated using the EBIS charge breeder to various experimental stations at ATLAS.

1.1 The ^{252}Cf source

The ^{252}Cf source ($t_{1/2}$ = 2.645 a; 3.1% fission branch) used in this system was prepared and provided by Oak Ridge National Laboratory. The ^{252}Cf is produced by successive neutron capture on an actinide target (originally primarily plutonium) using the High Flux Isotope Reactor, one of the highest flux reactors in the world [2]. The material is electroplated onto a stainless steel plate providing a fission fragment source for CARIBU. The current source is 500 mCi in total activity but effectively 130 mCi for radioactive recoils into the gas catcher due to the thickness of the deposit.

1.2 Helium gas catcher

The helium gas catcher is a large cylindrical vessel 50 cm in diameter and 120 cm in length with one end tapering into a nozzle for extraction of the fission-produced ions. The large stopping volume of the gas catcher reduces the residual energy of the fission products after they pass through the cover foil and degrader foil (as described below). With their energies greatly reduced, interactions with the helium in the gas catcher lead to final ionization states of 1^+ or 2^+ for the collected fission fragments. These ions are directed forward with a DC electric field and kept centered in the cell by parallel rings of electrodes with alternating (RF) potential until the gas flow pushes the ions out of the extraction nozzle.

1.3 Degrader assembly system

Prior to entering the gas catcher, the ions first pass through a thin gold cover to contain self-sputtering recoil emissions and prevent any flaking of the californium source into the gas catcher. Next, the ions pass through the aluminum degrader foil where much of their residual energy is reduced prior to entering the helium gas catcher.

In the past, a less-than-ideal cylindrical-shaped degrader was utilized due to production limitations. This resulted in non-uniform energy loss as the ions passed through the degrader. With the advent of 3D printing, a new hemispherical degrader was prepared to enable a more uniform energy loss.

* Corresponding author: mgott@anl.gov

 https://doi.org/10.1051/epjconf/202022906002

2 Degrader assembly preparation

All of the gas catcher system including the degrader assembly was built and handled to ultra-high vacuum standards as to not affect the purity of the gas; the helium gas is high purity and all contaminants must be kept at the ppb level. The introduction of the new degrader foil presents a possible source of contamination so the use of outgassing materials of any kind is prohibited. The design is such that no epoxies or cements are necessary for the preparation of the degrader assembly. Additionally, the whole setup was assembled and handled in a Class 100 clean bench.

2.1. Previous degrader assembly

The previous degrader [3] was cylindrical-shaped and was approximately 10 cm in diameter and 5 cm tall. An aluminum base and a thin aluminum ring connected by three thin aluminum posts providing the support structure for the thin aluminum foil. The posts also served to hold and clamp the aluminum degrader foil in place around the circumference of the ring. The thin ring on top held the top degrader foil in place. Figure 1 shows the completed degrader assembly with the aluminum foils installed.

Fig. 1. Previous generation of degrader foil.

2.2 New degrader design

The new degrader is a hemispherical design composed of six triangular pieces that screw into a heavy base ring of Inconel steel. Each of the triangular pieces is 3D printed to engineered specifications so that they piece together tightly. The aluminum foil is held in place by a metal gasket that screws into the triangular framing.

The pieces are all cleaned using an acid wash, water rinse, and alcohol rinse. The cleaned pieces are handled with gloves to prevent introduction of contaminants to the surface. All tools are cleaned with alcohol prior to use with this assembly.

2.3 Assembly of the degrader

2.3.1 Preparation of the Aluminum foils

High purity aluminum foil was purchased from Goodfellow (Coraopolis, Pennsylvania) with an areal density of 1.7 $mg{\cdot}cm^{-2}$. The foils were placed between two pieces of weighing paper to prevent tearing as the foils were cut. An in-house fabricated stencil was used to mark the foil to be cut (Figure 2). The foils were carefully cut across the marked lines to create the six foils necessary to fabricate the degrader.

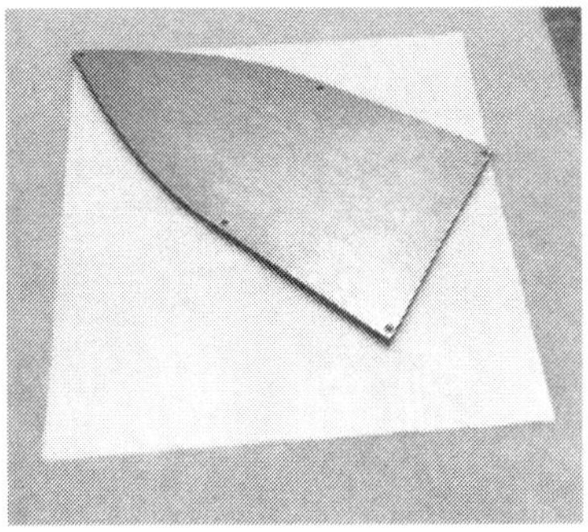

Fig. 2. Stencil placed on top of foil inside of weighing paper.

2.3.2 Piecing together the sections of the degrader

The cut foil is placed on top of the 3D-printed frame and, once positioned, the stainless steel gasket is carefully placed on top (Figure 3). A pick is used to punch screw holes in the foil at the base of the frame and then small (sized) screws are used to hold the foil and gasket in place. The gasket is bent around the frame and the screw at the top is attached next following the same procedure. Finally, the two screws on the side are put into place. As these final screws are tightened, the foil stretches slightly and is held tightly in place. The same procedure is followed for all six sections and great care must be taken not to damage/puncture the foil while putting the screws in place.

Fig. 3. Placement of foil on frame prior to addition of gasket.

2.3.3 Piecing together the degrader assembly

With the sections completed, the sections were each attached to the base plate using four vented screws. Initially the fragments were screwed in loosely to allow them flexibility to slide in beside one another. Once all sections were installed, a screw was installed at the top to hold them together at the same height and then all the screws at the base were tightened. The final completed degrader is shown in Figure 4.

Fig. 4. Completely assembled degrader.

3 Summary

The Californium Rare Ion Breeder Upgrade project provides neutron-rich beams for a variety of nuclear physics applications. A new, hemispherical degrader assembly has been prepared to provide a more uniform energy loss for the fission fragments as they enter the helium gas catcher. Care must be taken to avoid damage or contamination during assembly to prevent issues with degrader in the ultra-high vacuum environment in the system.

This material is based upon work supported by the U.S. Department of Energy, Office of Science, Office of Nuclear Physics, under Contract Number DE-AC02-06CH11357. This research used resources of ANL's ATLAS facility, which is a DOE Office of Science User Facility.

References

1. G. Savard, R.C. Pardo, S. Baker, C.N. Davis et al., Hyperfine Interact. **199**, 301 (2011)
2. R.C. Martin, J.B. Knauer, P.A. Balo, Appl. Radiat. Isot. **53**, 785 (2000)
3. J.P. Greene, G. Savard, R.C. Pardo, S. Baker et al., Nucl. Instr. Meth. Phys. Res. A **655**, 21 (2011)

https://doi.org/10.1051/epjconf/202022906003

Medical isotope collection from ISAC targets

Peter Kunz[1,2,*], *Corina* Andreoiu[2], *Victoria* Brown[1], *Marla* Cervantes[1], *Julia* Even[5], *Fatima H.* Garcia[2], *Alexander* Gottberg[1], *Jens* Lassen[1], *Valery* Radchenko[1,4], *Caterina F.* Ramogida[1,2], *Andrew K. H.* Robertson[1,3], *Paul* Schaffer[1,2], and *Rozhannaa* Sothilingam[1]

[1]TRIUMF, 4004 Wesbrook Mall, Vancouver, BC V6T 2A3, Canada
[2]Department of Chemistry, Simon Fraser University, Burnaby, BC V5A 1S6, Canada
[3]Physics and Astronomy, University of British Columbia , Vancouver, BC, V6T 1Z4, Canada
[4]Department of Chemistry, University of British Columbia , Vancouver, BC, V6T 1Z1, Canada
[5]University of Groningen, Zernikelaan 25, 9747 AA Groningen, Netherlands

Abstract. The ISAC facility (Isotope Separation and Acceleration) at TRIUMF has recently started to provide isotopes for pre-clinical nuclear medicine studies. By irradiating ISOL (Isotope Separation OnLine) targets with a 480 MeV proton beam from the TRIUMF H- cyclotron, the facility can deliver a large variety of radioactive isotope beams (RIB) for research in the fields of nuclear astrophysics, nuclear structure and material science with half-lives down to a few milliseconds via an electrostatic beamline network. For the collection of medical isotopes, typically with half-lives in the range of hours or days, we have developed a compact apparatus for the implantation of mass-separated RIB on a target disc at energies between 20-55 keV. In this paper, we also discuss two different retrieval methods of the implanted activity from the implantation target: by chemical etching of the target surface and by recoil collection of implanted alpha emitters.

1 Introduction

The ISAC facility features two target stations for the irradiation of ISOL-type thick targets with up to 100 μA of a 480 MeV proton beam from the TRIUMF H- cyclotron. Using a variety of refractory target materials (metal foils, composite ceramic carbides, oxides) a large number of radioactive isotope beams (RIB) can be produced by spallation, fragmentation or fission from elements such as C, Si, Ti, Ni, Nb, Ta, Th and U. Only sufficiently volatile isotopes in atomic or molecular form for which the average release time doesn't exceed their half-life too much can be released, ionized, mass-separated and distributed through the ISAC electrostatic beamline network. Therefore, to accommodate a fast and efficient release, ISAC targets are generally operated at temperatures as high as permissible by the properties of the target material – mainly vapor pressure, thermal conductivity, sintering behavior and chemical stability. Operating temperatures range from 1100 °C for nickel oxide targets to 2300 °C for tantalum metal foil targets [1]. The facility is mainly used for research in the fields of nuclear astrophysics, nuclear structure and material science [2]. In addition to these applications, the facility is now also producing radioisotopes for pre-clinical nuclear medicine research. Some of those isotopes are of interest for imaging or therapeutic applications or both (*theranostic* isotopes). They generally have half-lives in the range of hours or days and can be produced in sufficient quantities for pre-clinical research from a variety of target materials. The current status of verified RIB yields is available for ISAC users in the ISAC Yield Database [3].

This paper is providing an overview of of isotopes that have already been collected as well as potential candidates for future applications related to nuclear medicine research. In particular, the collection of the heavy, alpha emitting isotope ^{225}Ac and its precursor ^{225}Ra are discussed. ^{225}Ac [4] and its decay product ^{213}Bi are promising candidates for targeted alpha therapy (TAT). Furthermore, methods to transfer the collected activity from the implantation target have been investigated. Sample preparation methods such as chemical etching of implantation targets and the recoil collection of ^{213}Bi from the ^{225}Ra/Ac decay chain are discussed.

2 Experimental

The two target stations at the ISAC facility are operated alternately. The 480 MeV proton beam can be directed at either one of them where it impinges on a production target with a beam current of up to 100 μA with a maximum achievable beam power of 48 kW. In order to accommodate the full capacity of the cyclotron, high-power tantalum target containers [5] have been developed. The addition of cooling fins increases the emissivity to 92% and allows to dissipate up to 25 kW of beam power. This is sufficient since power deposition in the target material is usually less than 50% of the maximum beam power. Thickness, density, thermal conductivity, vapor pressure are fac-

*e-mail: pkunz@triumf.ca

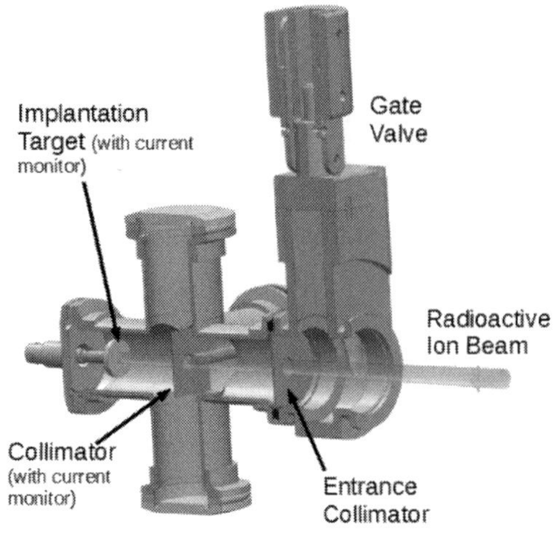

Fig. 1. Section view of the ISAC Implantation chamber. The RIB is focused though a set of collimators on the implantation target in the back of the chamber.

tors that determine the maximum feasible beam power on target. The extraction of a beam from the ion source is only efficient under high-vacuum conditions ($< 10^{-5}$ mbar) [6]. The vapor pressure depends strongly on the target temperature which depends on power deposition and the efficiency of power dissipation, governed by the thermal conductivity of the target material and the emissivity of the target container. The resulting target temperature limitations can be detrimental to the efficient and fast release of isotopes through diffusion and effusion which benefits from high operating temperatures. The best balance between these two opposing requirements can be achieved by using refractory materials with high thermal stability, low vapor pressure and good thermal conductivity. Another aspect considering target material properties are diffusion times. Slow diffusion reduces the release efficiency of short-lived isotopes. This effect can be mitigated by using high-porosity materials. For example, the average bulk density of the uranium carbide used for ISAC targets is typically around 3.3 g/cm^3 [7], about 1/3 of the theoretical density of UC_2. In comparison to bulk UC_2, the thermal conductivity of this porous material is reduced by a factor of 6. In order to still achieve sufficient power dissipation, composite ceramic target materials [8], consisting of disks made of a 100-250 μm thick carbide layer on a 130 μm graphite backing foil have been developed. This configuration increases the overall thermal conductivity significantly due to the relatively high conductivity of the graphite backing foils. For low-thermal conductivity oxide materials (NiO, UO_2) 25 μm thick niobium metal backing foils serve as thermal conductivity booster.

The production rate for a specific isotope is the product of production cross section, proton beam intensity and target thickness. The actual isotope beam intensity that arrives at the experiment also depends on the release efficiency form the target, the ionization efficiency and the transport efficiency of the ion beam through the ISAC electrostatic beamline network. Beam transport includes the extraction and acceleration of ions from the ion source to energies between 10-55 keV, subsequent mass separation through a low resolution pre-separator, a high-resolution mass separator (m/Δm=2000).

For the ionization process a variety of ion source options can be utilized [9]. Which ionization method is most suitable, depends strongly on the ionization potential of the atoms or molecules released from the target. The ion beams discussed in this paper were ionized with a surface ion source or by resonant laser ionization.

For the production of heavy isotopes for targeted alpha therapy research, composite ceramic uranium carbide targets [7] are currently used. They have typical thicknesses between 0.05 and 0.1 mol U/cm^2 and are irradiated with a proton beam of max. 10 μA for up to 5000 μAh (current operation license limits) at an operating temperature of approximately 1950 °C. Neutron-deficient isotopes for medical applications in the lanthanide region are efficiently produced from tantalum metal foil targets, which are filled with several hundred 25 μm thick Ta foil disks up to a total thickness of 0.14 mol/cm^2. They are operated at up to 70 μA proton beam current and a target core temperature of 2300°C [6].

2.1 ISAC Implantation Station (IIS)

The IIS is located a few meters behind the ISAC high-resolution mass separator on a short branch of the electrostatic beamline. It is divided in two sections. Section 1 contains ion beam optics for beam positioning, focusing or rastering. It ends at a gate valve with a pumping port on it's downstream side which provides vacuum for the ion collection chamber (section 2). This chamber can be custom-build for specific applications and is attached to the gate valve by the user.

The design of the ion collection chamber used for the examples presented in this paper is shown in figure 1. It is made of a standard KF40 6-way cross. Attached are vent and sampling valves to test for airborne contamination, a vacuum gauge, a collimator with current readout for beam centering and the implantation target mounted on an electrical feedthrough to allow current monitoring during the implantation process. In principle the target can have any form and composition. The only condition is that it must be electrically conductive to avoid charging and enable a current readback. For the experiments in this paper we used a standard aluminium SEM sample holder (⌀25 mm) with a machined-in 2 mm deep cut-out (⌀16 mm), see figure 2. The cut-out is required for the etching process described in 2.2. A manual KF40 gate valve serves as an interface to section 1. Once an implantation is finished the gate valves in both sections are closed and the volume between them is vented. Now, the ion collection chamber can be detached and - still under vacuum - be transferred to the radiochemistry laboratory. This way, the collected activity can be considered a closed source, which is important particularly for alpha emitters where the spread of contamination through recoil emission can be a problem.

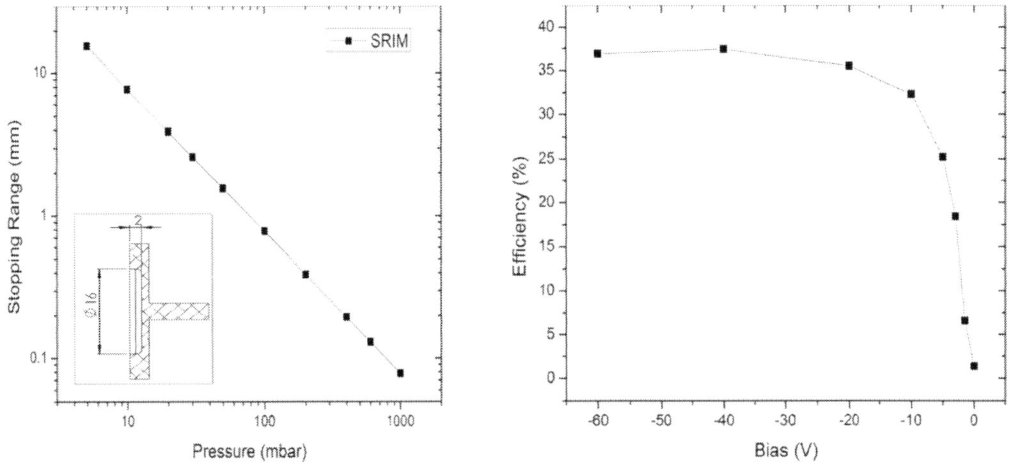

Fig. 2. ^{225}Ac recoil collection depends on the stopping range of approx. 100 keV alpha decay recoils (left) and the bias between the implantation target (insert, dimensions in mm) and an identical ion catcher target. The measured ^{213}Bi collection efficiency (right) at normal air pressure for a configuration in which the catcher target was pressed against the implantation target, separated by a thin insulator. A negative bias was applied to the catcher.

Once in the lab, the chamber can be vented and the implantation target can be retrieved for further processing.

2.2 Retrieval of activity from IIS targets

We have explored 2 methods to retrieve radioactive isotopes from the implantation target: chemical etching of the target surface (2.2.1) and recoil collection from implanted alpha emitters (2.2.2)

2.2.1 Chemical etching

The implantation depth of a heavy ^{225}Ac/Ra ion beam at a typical ISAC extraction energy of 30 keV in aluminium is about 20 nm [10]. Therefore, the transfer of the implanted isotopes into an aqueous solution requires the removal of at least 20 nm of the implantation target surface.

To achieve this, 0.1 ml of 0.1N HCl is filled into the implantation target cut-out and then gently evaporated on a hot plate. Then the target is cooled down and a thin layer of $AlCl_3$, that has formed during the evaporation, is dissolved with 0.1 ml 0.1N HCl. This solution is picked up with a micropipette and transferred to a vial. This etch-rinse procedure is repeated several times. 3 iterations already remove >95% of the implanted activity, leaving it in a small volume of acidic solution which can be used for further radiochemical experiments.

2.2.2 Recoil collection

Isotopes like $^{223-225}$Ra and ^{225}Ac decay through relatively long chains, including multiple alpha decays. The recoil energy transferred during an alpha decay to the residual nucleus is around 100 keV, about 3 times larger than the typical ion beam implantation energy at ISAC. Therefore, the alpha decay recoils have sufficient kinetic energy to escape as long as the recoil vector points out of the target. For a thick implantation target, as was used in these experiments, this leads to a theoretical release efficiency of less than 50 % for a single alpha decay. However, the collection of ^{213}Bi from a ^{225}Ac source involves 3 alpha decays in the chain. The decays of ^{221}Fr and ^{217}At contribute to the overall release efficiency, increasing it to 60% at an implantation energy of 30 keV. This result was confirmed with a GEANT4 [11] recoil track simulation.

Using a ^{225}Ac/Ra target as shown in figure 2, ^{213}Bi recoils were accumulated by placing an identical collection target, separated by a thin insulator, directly on top. The collection was ended when the activity of ^{213}Bi ($T_{1/2}$ = 45.59 min) had reached saturation. Figure 2 (right) shows that a negative bias of >20 V lead to a collection efficiency for ^{213}Bi of ~35 %.

The stopping range of recoils in air under normal pressure is less than 0.1 mm, see figure 2 (left). Aligning the stopping range more accurately with the distance between source and collector by reducing the pressure to 50 mbar and using argon as a buffer gas increased the collection efficiency to 47%. However, for applications such as chelation studies a 35% efficiency was sufficient and sample collection under normal air pressure much more convenient.

3 Conclusion

All measured yield rates from ISAC targets are compiled in Ref. [3]. From these values, activities can be extrapolated either until saturation is reached or to a maximum 24 hour implantation period. This is a realistic time frame, considering that radioactive ion beam time is usually allocated for several experiments with one ISAC target. Within this period, a ^{225}Ac beam ($T_{1/2}$ = 9.92 d) at a rate of $1.3 \cdot 10^8$ ions/s from a uranium carbide target / laser

Table 1. Selected isotopes for potential medical applications from ISAC targets. The table lists primary collection beam, target material, ion source and established yields [3] and the activity that can be accumulated within a typical 24h beam collection period.

Isotope	Half-life	Application	Primary beam	Target material	Ion source	Measured yield [ions/s]	Accumulated 24h Activity [Bq]
225Ac	9.92 d	TAT^a	225Ac	UC	LIS^{ff}	1.30E+08	8.60E+06
225Ac	9.92 d	TAT^a	225Ra	UC	SIS^{gg}	1.70E+08	3.40E+06
224Ra	3.66 d	TAT^a	224Ra	UC	SIS^g	1.60E+09	2.70E+08
223Ra	11.43 d	TAT^a	223Ra	ThO	SIS^g	6.90E+08	4.00E+07
213Bi	45.6 m	TAT^a	225Ac	UC	SIS^g	6.70E+07	2.25E+06
212Pb	10.64 h	TAT^a	224Ra	UC	SIS^g	1.60E+09	1.00E+08
212Bi	60.6 m	TAT^a	224Ra	UC	SIS^g	1.60E+09	7.80E+07
211At	7.22 h	TAT^a	211Rn	UO	FEBIAD^h	1.00E+08	2.40E+07
211At	7.22 h	TAT^a	211At	UO	FEBIAD^h	8.40E+07	7.40E+07
211At	7.22 h	TAT^a	211Fr	UC	SIS^g	1.90E+09	4.30E+07
209At	5.4 h	SPECT^b	213Fr	UC	SIS^g	1.80E+09	8.50E+08
209At	5.4 h	SPECT^b	209At	ThO	LIS^f	1.90E+08	1.70E+08
177Lu	6.65 d	BT^c	177Lu	Ta	LIS^f	6.50E+08	6.40E+07
169Yb	32.02 d	SPECT^b	169Yb	Ta	LIS^f	5.10E+10	1.00E+09
166Yb	2.36 d	AT^d	166Yb	Ta	LIS^f	1.50E+11	3.70E+10
165Er	10.3 h	AT^d	165Yb/165Tm	Ta	LIS^f	9.32E+10	3.90E+10
161Ho	2.5 h	AT^d	161Ho	Ta	LIS^f	9.96E+09	9.30E+09
161Ho	2.5 h	AT^d	161Er	Ta	LIS^f	1.03E+10	3.90E+09
149Tb	4.12 h	TAT^a/SPECT^b	149Tb	Ta	SIS^g	5.78E+08	5.75E+08
140Nd	3.37 d	PET^e/AT^c	140Sm	Ta	LIS^f	9.70E+08	1.70E+08
124I	4.18 d	PET^e	124I	UO	FEBIAD^h	2.20E+08	3.20E+07
123I	13.22 h	PET^e	123I	UO	FEBIAD^h	3.80E+07	2.60E+07
82Sr	1.35 d	PET^e	82Sr	Nb	SIS^g	1.00E+10	2.10E+08
77Br	57.0 h	Labeling	77Rb	Nb	SIS^g	1.60E+09	3.70E+08
67Ga	78.28 h	Imaging	67Ga	ZrC	LIS^f	8.10E+09	1.50E+09
67Cu	2.58 d	BT^c	67Cu	Ta	LIS^f	1.40E+08	3.10E+07
64Cu	12.70 h	PET^e/therapy	64Cu	Nb	LIS^f	3.20E+09	2.20E+09
61Cu	3.33 h	PET^e	61Cu	Nb	LIS^f	7.80E+08	7.50E+08

^{a} targeted alpha therapy, ^{b} single-photon emission computed tomography, ^{c} beta therapy, ^{d} Auger therapy, ^{e} positron emission tomography, ^{f} laser ion source, , ^{g} surface ion source , ^{h} forced electron beam induced arc discharge ion source

ion source combination accumulates an activity of up to $8.6 \cdot 10^6$ Bq. The accompanying ^{225}Ra ($T_{1/2}$ = 14.8 d) beam is collected simultaneously. At a rate of $1.7 \cdot 10^8$ ions/s from a uranium carbide target / surface ion source combination a total of $1.4 \cdot 10^{13}$ atoms ($7.5 \cdot 10^6$ Bq) can be collected in 24 hours. ^{225}Ra decays via β-decay into ^{225}Ac, reaching a maximum activity of $3.4 \cdot 10^6$ Bq ^{225}Ac approximately 18 days after the implantation.

The etching method (section 2.2.1) was used to extract Ac and Ra isotopes. Then the radium was separated from the actinium by well established extraction chromatography methods [12]. The remaining ^{225}Ra can be recycled and used as an ^{225}Ac generator. This has been demonstrated as part of the development of Multi-isotope SPECT imaging of the ^{225}Ac decay chain by Robertson et. al. [13]. The collection of ^{213}Bi via recoil transfer was employed during chelation studies. In this case the collected activities were significantly lower, in the kBq range, but still sufficient for this kind of experiment. The recoil transfer from the low energy beta decay of ^{225}Ra is insignificant compared to the implantation energy and, therefore, stays in the implantation target, while it's decay product ^{225}Ac can escape, thus, extending the lifetime of the ^{213}Bi source significantly.

With the example of ^{225}Ac/Ra beams, it has been demonstrated that the ISAC facility at TRIUMF has the capability of providing pure isotope samples for pre-clinical nuclear medicine research [14]. Infrastructure for the ISAC Implantation Station and methods for production and retrieval of medical isotopes have been established. Similar implantation procedures have also been applied to the production of 209/211At via the implantation of $^{211/213}$Fr beams for At-based α-therapy research [15][16]. In principle, the methods described above are applicable for any isotope beam from ISAC targets.

Table 1 lists the established yield rates (Ref. [3]) of isotopes of potential interest for nuclear medicine research from various targets and ion sources. Based on the established yields, the activity accumulated over a 24h implantation period was extrapolated. The data shows that in particular for some lanthanides very high activities can be collected. However, for relatively long-lived isotopes, such as ^{225}Ra and ^{225}Ac larger activities could be obtained by simply increasing the implantation period.

TRIUMF receives federal funding via a contribution agreement through the National Research Council of Canada. We also acknowledge additional support through Discovery Grants from the Natural Sciences and Engineering Research Council of Canada (NSERC): SAPIN/00021/2014 (P. Kunz) and SAPIN/00039/2017 (J. Lassen). Thanks to the ISAC operations crew for their support

References

[1] M. Dombsky, P. Kunz, Hyperfine Interactions **225**, 17 (2013)

[2] J. Dilling, *ISAC and ARIEL: The TRIUMF radioactive beam facilities and the scientific program*

(Springer, New York, 2014), ISBN 978-94-007-7962-4

[3] P. Kunz, *ISAC Yield Database - https://mis.triumf.ca/science/planning/yield/beam* (2018), `https://mis.triumf.ca/science/planning/yield/beam`

[4] A.K.H. Robertson, C.F. Ramogida, P. Schaffer, V. Radchenko, Current Radiopharmaceuticals **11**, 156 (2018)

[5] P. Bricault, M. Dombsky, A. Dowling, M. Lane, Nuclear Instruments and Methods in Physics Research Section B: Beam Interactions with Materials and Atoms **204**, 319 (2003)

[6] M. Dombsky, P. Bricault, V. Hanemaayer, Nucl. Phys. A **746**, 32c (2004)

[7] P. Kunz, P. Bricault, M. Dombsky, N. Erdmann, V. Hanemaayer, J. Wong, K. Lützenkirchen, Journal of Nuclear Materials **440**, 110 (2013)

[8] V. Hanemaayer, P. Bricault, M. Dombsky, Nuclear Instruments and Methods in Physics Research Section B: Beam Interactions with Materials and Atoms **266**, 4334 (2008)

[9] P.G. Bricault, F. Ames, M. Dombsky, P. Kunz, J. Lassen, A. Mjøs, J. Wong, Nuclear Instruments and Methods in Physics Research Section B: Beam Interactions with Materials and Atoms **366**, 34 (2016)

[10] J.F. Ziegler, M. Ziegler, J. Biersack, Nuclear Instruments and Methods in Physics Research Section B: Beam Interactions with Materials and Atoms **268**, 1818 (2010)

[11] J. Allison, K. Amako, J. Apostolakis, P. Arce, M. Asai, T. Aso, E. Bagli, A. Bagulya, S. Banerjee, G. Barrand et al., Nuclear Instruments and Methods in Physics Research Section A: Accelerators, Spectrometers, Detectors and Associated Equipment **835**, 186 (2016)

[12] E.P. Horwitz, M.L. Dietz, R. Chiarizia, Journal of Radioanalytical and Nuclear Chemistry Articles **161**, 575 (1992)

[13] A.K.H. Robertson, C.F. Ramogida, C. Rodríguez-Rodríguez, S. Blinder, P. Kunz, V. Sossi, P. Schaffer, Physics in Medicine and Biology **62**, 4406 (2017)

[14] N.A. Thiele, V. Brown, J.M. Kelly, A. Amor-Coarasa, U. Jermilova, S.N. MacMillan, A. Nikolopoulou, S. Ponnala, C.F. Ramogida, A.K.H. Robertson et al., Angewandte Chemie International Edition **56**, 14712 (2017)

[15] J.R. Crawford, P. Kunz, H. Yang, P. Schaffer, T.J. Ruth, Applied Radiation and Isotopes **122**, 222 (2017)

[16] J.R. Crawford, H. Yang, P. Kunz, D.S. Wilbur, P. Schaffer, T.J. Ruth, Nuclear Medicine and Biology **48**, 31 (2017)

Calcium targets for production of the medical Sc radioisotopes in reactions with p, d or α projectiles

Anna Stolarz[1*], *Mateusz* Sitarz[1,2], *Katarzyna* Szkliniarz[3], *Jarosław* Choiński[1], *Jerzy* Jastrzębski[‡1], *Agnieszka* Trzcińska[1], and *Wiktor* Zipper[3]

[1]Heavy Ion Laboratory, University of Warsaw, Warsaw, Poland
[2]Faculty of Physics, University of Warsaw, Warsaw, Poland
[3]Institute of Physics, University of Silesia, Katowice, Poland
‡ deceased

Abstract. The scandium radioisotopes for medical application can be produced in reactions of calcium with proton, deuteron or alpha projectiles. Enriched isotopic calcium material is commercially available mainly as calcium carbonate which can be used directly for production of Sc radioisotopes or can be converted into other calcium compounds or into metallic form. The superiority of application of calcium oxide is shown throughout analysis of use of each target chemical form.

1 Introduction

Majority of medically interested radioisotopes are produced in reactions induced by neutrons i.e. in reactors. Nevertheless, the alternative methods of their production are being extensively developed. The advantages and drawbacks of each production route are well presented by M. A. Synowiecki et al in [1].

The studies on the alternative methods were triggered by unplanned shut-downs of reactors several years ago which caused the shortage of isotopes (i.e. ^{99}Tc) widely used in medical applications. These studies are also stimulated by development of the diagnostic techniques and by search for replacements having longer half-live then isotopes recently used in PET scanning. One of the alternative method of these radioisotopes production is use of the reactions induced by accelerated projectiles such as protons, deuterons or alpha particles.

These studies cover search for replacements of PET isotopes such like ^{18}F ($T_{1/2}$≈110 min), ^{68}Ga ($T_{1/2}$≈68 min, ^{60}Cu ($T_{1/2}$≈24 min), ^{61}Cu ($T_{1/2}$≈20 min) and ^{15}O ($T_{1/2}$≈122 s). The scandium isotopes are among potential replacements. The scandium radioisotopes which are β^+ emitters (^{44g}Sc and ^{43}Sc, $T_{1/2}$=3.97 h and 3.89 h, respectively) can replace currently used PET scanning isotopes mainly due to their longer half-life. Another interesting scandium isotope, ^{47}Sc ($T_{1/2}$=3.35 d) with its low energy β^- emission, is a good candidate for targeted radiotherapy. This isotope coupled with one of the Sc β^+ emitters may represent an ideal theranostic pair. The ^{44}Sc is as well considered a good candidate to be applied in the new β^+-γ coincidence PET method [2].

Majority of the studies on the scandium isotopes production via reactions induced by p, d or α projectiles are performed using calcium as a target nucleus. The alternative nucleus is Ti as the source of Sc isotopes but comparison of the cross sections for reactions of both nuclei (Table 1) shows that reaction with Ca nucleus promises much higher production efficiency. As can be found from these data production of ^{43}Sc with very high efficiency can be achieved even applying target composed of the natural calcium employing reaction of ^{40}Ca (96.94 % of nat. abundance) with α particles [3].

Table 1. Example of cross sections for production of ^{43}Sc and ^{44g}Sc on reactions of Ca or Ti with p, d or α [4].

isotope produced	Calcium isotope cross section [mb]/ (beam energy [MeV])			Titanium isotope cross section [mb]/ (beam energy [MeV])	
	p	d	α	p	d
^{43}Sc	^{43}Ca 400/(9) ^{44}Ca 200/(24)	^{42}Ca 270/(5)	^{40}Ca 700/(15)	^{46}Ti 60/(15) 80/(60) ^{47}Ti 70/(25)	
^{44g}Sc	^{44}Ca 700/(10)	^{43}Ca 380/(5)	^{42}Ca 750/(25) ^{43}Ca 600/(35)	^{46}Ti 240/(35) ^{47}Ti 70/(15) ^{48}Ti 80/(25)	^{46}Ti 110/(10); 310/(75) ^{47}Ti 140/(22); 200/(100)

* Corresponding author: anna@slcj.uw.edu.pl

Table 2. Advantages (+) and drawbacks (-) of use of various chemical form of calcium targets.

Chemical form	$CaCO_3$	CaO	Ca
	(+) The form of the enriched materials easily available commercially; (+) Ready to be used for target preparation by encapsulation, pellets formation, etc.	(+) Achievable by thermal decomposition of $CaCO_3$; (-) Unstable in air , quick manipulation is required or best to be handled in the inert atmosphere.	(-) Have to be prepared by $CaCO_3$ conversion; it is a two steps procedure; (-) unstable in the air, it requires handling in the inert atmosphere.
under the beam	(-) Thermal insulator; Decomposes producing $CaO+CO_2$. The target cracks when exposed to intensive beams due to this process (see Fig. 1); (-) Production of large amount of ^{13}N (decaying in ~10 min via β^+ to ^{13}C).	(-) Thermal insulator; (+) Low production of ^{13}N.	(-) If not sufficient cooling can melt in the beam.
Treatment after irradiation	(+) Dissolves very easily in a weak acids.	(+) Dissolves easily in a weak acids; only slightly more difficult than carbonate.	(++) Dissolves even in water.
Efficiency	see next table		

2 Chemical form of the target

Production of the Sc medical radioisotopes in reaction of Ca nucleus can be done working with unprocessed enriched material i.e. with calcium carbonate ($CaCO_3$, the chemical form of the enriched Ca isotopes mostly available commercially) or with material converted into either calcium oxide (CaO) or metal (Ca). Work with each target form has advantages and drawbacks (Table 2). The thermal damages mentioned in Table 2 for calcium carbonate can be eliminated by mixing the target material with a good heat conductor e.g. graphite or aluminium as discussed in [5].

Fig. 1. $CaCO_3$ target after 2 h irradiation with 15 μA proton beam of 16 MeV.

Taking into account only the level of produced activity which is proportional to the nuclei number per cm² as expressed in the formula

$$A = \Phi \, n \, \chi \, \sigma \, (1-e^{-\lambda t}) \quad (1)$$

where:

A - activity [Bq]
Φ - beam intensity [projectiles/s]
n - number of nuclei per mg of the target material
χ - target thickness [mg/cm²]
σ - reaction cross section [cm²]
λ - decay constant ($=\ln2/T_{1/2}$) [s^{-1}]
t - irradiation time [s]

working with metallic form of Ca would be the most favourable. The activity of the Sc radioisotopes produced during the same time of irradiation of Ca metallic target would be nearly tripled comparing to activity produced while using $CaCO_3$ (or doubled if working with CaO). This is due to the number of nuclei per cm² (N) in targets with thicknesses covering the projectile range in $CaCO_3$, CaO and Ca. The ratio of nuclei numbers in these targets is ~ 1:2:3, respectively, and thus is the produced activity.

Table 3. 16 MeV proton and 25 MeV alpha ranges and stopping powers in calcium targets

Chemical form, density	Range* of projectile in material [mg/cm²]		Stopping power* MeV/(mg/cm²)		N [nuclei/cm²]	
	16 MeV p	25-8 MeV α**	16 MeV p	25 MeV α	16 MeV p	25 MeV α
$CaCO_3$ 2.71 g/cm³	348.59	56.94	0.02576	0.2157	$3.48\times A_v\times10^{-3}$	$0.57\times A_v\times10^{-3}$
CaO 3.35 g/cm³	368.50	72.025	0.02450	0.2028	$6.58\times A_v\times10^{-3}$	$1.28\times A_v\times10^{-3}$
Ca 1.55 g/cm³	389.05	77.068	0.02345	0.1922	$9.72\times A_v\times10^{-3}$	$1.92\times A_v\times10^{-3}$

Av – Avogadro constant = 6.022140857×10^{23} mol⁻¹;
* calculated using SRIM 2013 code;
** example energy for ^{43}Sc isotope production in ^{40}Ca reaction with α;

However, conversion of $CaCO_3$ into metallic Ca is a time-consuming two steps reduction process [6,7]. The process can be done under vacuum by decomposing carbonate to oxide, followed by the oxide reduction into Ca using metallic (Me) reductants such as e.g. Zr, Ti.

$CaCO_3 \Rightarrow CaO + CO_2$ (heating at temp. > 700 °C) (2)

$2CaO + Me \Rightarrow Ca + MeO$ (reduction) (3)

In addition, process may as well introduce additional contaminants to the target apart from those present in the available starting material. Also the process efficiency (lower than 80%) has to be kept in mind considering metallic target. Therefore, it is better to avoid this conversion if it's not vital.

Working with metallic Ca would also require a special vacuum containers and/or construction of a transfer vacuum line to the cyclotron to prevent the contact of Ca with air.

Taking these difficulties into account it is much better to work with calcium oxide as target. Although activity produced is only doubled comparing to activity produced with $CaCO_3$ target of adequate thickness (see Table 3) but conversion to CaO is much easier than conversion to Ca. It can be done either by heating the oxide in flow of the inert gas [8] or in vacuum using the resistant heating. The advantages of the second method are: the instant/online control on the decomposition process via controlling the vacuum and gives the possibility of cooling down the produced CaO in the air free atmosphere. Conversion carried in a special vessel/crucible with the perforated cover (Fig. 2.) and venting-in the vacuum apparatus after completion of the procedure, with inert gas allows the transfer of the produced CaO to the glove box for manipulations needed to produce the final target (e.g. pressing the pellet, encapsulating into container, etc.) without special precautions.

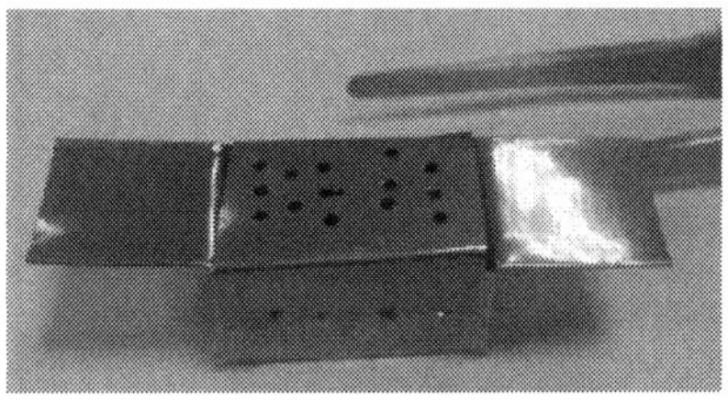

Fig. 2. Crucible with perforated cover used for calcium carbonate conversion in vacuum by resistant heating.

In addition decrease of the oxygen content in the target results in a significant decrease of the side radioactivity in the irradiation area related to the production of ^{13}N in $^{16}O(p,x)^{13}N$ or $^{16}O(d,x)^{13}N$ reaction.

The oxide targets prepared as inserts into graphite bed as described in [5] survived 45 min irradiation with 15 μA proton beam very well. There were no signs of thermal damage of the target. The irradiation conditions are sufficient to produce ~8 GBq of ^{44g}Sc irradiating the CaO converted from enriched up to 99.2 % $^{44}CaCO_3$. Taking into account the activity loses during isotope separation and labelling process, such amount of ^{44}Sc should be enough for diagnosing ~ 75 patients (the estimation based on clinical studies for ^{44}Sc [9] where 50.5 MBq of ^{44}Sc-PSMA-617 were applied for single diagnosis).

3 Conclusions

Production of the research quantities of Sc radioisotopes can easily be performed using targets made directly from calcium carbonate. However, for clinical application when higher activities are required more favourable is to work with targets made from calcium oxide.

Conversion of carbonate into oxide is one step process practically without loses of the often expensive, enriched calcium material.

As it has been shown using the calcium oxide instead of carbonate gives nearly double activity within the same time of irradiation and much less undesirable radioactivity (originating from decay of ^{13}N formed in the side reaction) is produced in the irradiation area.

Studies on Sc radioisotopes production presented in this paper were partly supported by NCBiR (National Centre for Research and Development), grant no. DZP/PBS3/2319/2014 and was partly performed within the framework of the EU HORIZON 2020 project RIA-ENSAR2.

References

1. Mateusz A. Synowiecki, L.R. Perk, J.Frank W. Nijsen, EJNMI Radiopharmacy and Chemistry, 3:3 (2018)
2. C. Lang, D. Habs, K. Parodi, P.G. Thirolf, Sub-millimeter nuclear medical imaging with reduced dose application in positron emission tomography using β γ coincidences. Preprint Typeset in JINST Style - HYPER VERSION (arXiv:1305.4261 [physics.med-ph]) (2013)
3. Katarzyna Szkliniarz, M. Sitarz, R. Walczak, J. Jastrzębski, A. Bilewicz, J. Choiński, A. Jakubowski, A. Majkowska, A. Stolarz, A. Trzcińska, W. Zipper, Appl. Radiat. Isot, **118**, 182-189 (2016)
4. N. Soppera, E. Dupont, M. Bossant, JANIS *Book of proton / deuteron / alpha-induced cross-sections. Comparison of evaluated and experimental data from JENDL/AN-2005, TENDL-2011 and EXFOR,* OECD NEA Data Bank, (June 2012)
5. Anna Stolarz, J. A. Kowalska, J. Jastrzębski, J. Choiński, M. Sitarz, K. Szkliniarz, A. Trzcińska, W. Zipper; AIP Conference Proceedings, **1962**, 020004 (2018); doi.org/10.1063/1.5035517
6. J.D. Stinson, Proceedings of the 1974 annual conference, (INTDS), 100 (1974)
7. G.E. Thomas, Report ANL/PHY/MSD-76-1, 255 (1976)

8. Nicholas van der Meulen, R. Hasler, C. Vermeulen, T. Corzaniga, S. Branccini, *Targetry developments for 44Sc production using enriched CaO*, 17th WTTC, Coimbra, Portugal, August 2018
9. Elisabeth Eppart, A. de la Fuente, M. Benešová, A. Khawar, R. A. Bundschuh, F. C. Gärtner, B. Kreppel, K. Kopka, M. Essler, F. Rösch, Theranostics, **7**, 4359 (2018)

AUTHOR INDEX